FOREWORD

We need another and a wiser and perhaps a more
mystical concept of animals. We patronize them for their
incompleteness, for their tragic fate of having taken form
so far below ourselves. And therein we err and err greatly.
For the animal shall not be measured by man. In a world
older and more complete than ours, they move finished
and complete, gifted with extensions of the senses we have
lost or never attained, living by voices we shall never hear.
They are not brethren; they are not underlings; they are
other nations, caught with ourselves in the net of life and
time, fellow prisoners of the splendor and travail of the
earth.

- Henry Beston.

It's understandable that the message in this book may be hard
to read for anybody who cares about animals, or for anyone who
cares about people too. I apologize for that, for having to describe
so many disturbing, sometimes heart-wrenching, details. The story
of wildlife and man's treatment of it is unfortunately a negative one.
There's no "positive" way to discuss the horrendous poaching of elephants in Africa, or the atrocities of lion, tiger, and bear "farms" in
Asia, or the sufferings and deaths of people from infectious diseases
that have increased due to climate change. But the first step in solv-
ing a problem is knowing that it exists, and hopefully, by describing

these problems and how they adversely affect both animals and people, some changes might be brought about and all of us will benefit as a result. The "good news" is that helping animals would be a very positive step toward helping people, as we will show in abundant examples. In the climate change issue, to take just one example, we can wrest positive results out of what would otherwise be very negative; if we can curtail global warming, we have the opportunity to save, literally, hundreds of millions of lives, both animal and human, over centuries of time.

But before we can get to the good news, how people's lives can be made safer, healthier, and better by changes in the ways we treat animals, we have to look at the ways that we treat them, and the story isn't pretty. There are some animals that are treated well by a few people, mostly dogs and cats and some horses, but for the most part humanity treats its non-human fellow travelers on the earth in terrible ways, from farm animals languishing on factory farms, to animals in rodeos and circuses, wildlife trying to escape from hunters or trappers, fish taken violently out of their natural habitat, animals driven to extinction, and on and on. But the good news is, at the risk of being repetitive, if we would change how we treat animals, people would benefit as much as the animals, including saving human lives and human suffering.

I had several reasons for writing this book. One is a lifelong desire to do whatever I can to help animals, and the second is the equal desire to help people, and especially to show people this connection between their own well-being and mankind's behavior toward animals. I reasoned that my experience as a scientist could be put to good use in doing the background research and organizing the data. I decided at the outset that I would be as objective as possible. My thought was that, even though pleas for kindness and respect for animals have been written about by many individuals for many years, it apparently hasn't worked, hasn't been enough, and animals

Helping Animals
Means Helping People

*From Hunting
and Poaching to
Climate Change
and Nuclear War*

HAROLD HOVEL

TABLE OF CONTENTS

Foreword i

Chapter 1: Helping Animals 1
   Sport Hunting 1
   Wildlife Killing Contests 7
   Pheasant Stocking 9
   Wildlife Services (formerly Animal Damage Control) 11
   Poaching 16
   Canned Hunts 27
   Wildlife Refuges 31
   Extinction 34

Chapter 2: Helping People 41
   Hunting Accidents, Injuries, Fatalities 41
   Lyme Disease 45
   Deer-Car Collisions 48
   Extinction 53
   Poaching, Park Rangers, and Terrorism 60
   Trapping 65

Chapter 3: Marine Wildlife 69

Chapter 4: Ocean Wildlife, Human Health, and Human Risk 78

Chapter 5: Climate Change 86
   Timeline 86
   Dynamics and Consequences of Climate Change 93

Temperature 97

Greenhouse Gases 101

Wildfires.. 104

Sea Level Rise.. 106

Extreme Weather. 109

Agriculture 113

Ethics.. 116

Helping People and Animals................................ 117

Diseases: Human Health and Animal Health................. 124

Diet and Climate Change 128

Energy 131

Natural or Man-Made 132

Summary 134

Chapter 6: Nuclear War.................................... 137

Chapter 7: A Few Suggestions 142

Epilogue 148

References... 152

Index 198

continue to be treated badly all over the world. Perhaps showing people that helping animals is really in their own selfish best interest, they would be more receptive to treating them well.

Remaining objective in describing so many areas where animals are being harmed, abused, and killed turned out to be no easy task, and when I came to the topic of hunting and its related topics (predator killing contests, government-sponsored wildlife destruction a.k.a. Wildlife Services, canned hunts, poaching, extinction), I found it impossible to remain objective, especially since many writers before me have pointed out that mankind in effect has an *ongoing war on wildlife*. All the research I carried out more than reinforced that statement.

The Christian and Jewish God, Yahweh, the Islamic God Allah, the Great Spirit, Budda, and probably others in other Beliefs have all been said to have created the Earth and the animals and people on it. How can we humans who believe in these Faiths assume that we have the right to wantonly destroy what They created, to treat other species with thoughtless or deliberate cruelty, to deny their needs and comforts, to kill them every year in the billions for our own wants and desires?

There are a number of animal-related issues where the connection between helping animals and helping people isn't obvious, except in the *moral and humanitarian* sense. Some of these issues are genuinely egregious and horrifying. One example is the "crush videos" wherein scantily-clad women beat, stomp-upon, and torture to death small animals, usually kittens and puppies, selling the video recordings to people that apparently derive sexual pleasure out of watching them. Other, perhaps less egregious issues but not much less, might include animals in entertainment where chimps are beaten to make them perform, horses are tortured by acid placed on their legs to cause them to "high-step" in an attempt to minimize their pain, and dog racing where greyhounds are trained by chasing a chained rabbit and otherwise lead lives of solitude and caged misery between races. Horse racing is another topic, where horses are forced to race at young ages where their bones haven't fully formed and many hundreds die each year from broken bones or other injuries during races, while the horses that don't win or who stop winning are often sent to the pain and terror of the slaughterhouse.

In other topics, the benefit to people may exist but may seem marginal. The treatment of circus animals, especially elephants, comes in this category. Elephants are usually captured as infants or adolescents, beaten repeatedly to break their spirit while teaching them tricks, chained in one place for most hours of the day between circus acts (elephants in the wild wander many miles during the day), and abused with the infamous bullhooks with which they are jabbed, grabbed in sensitive body areas, and sometimes beaten. Upon occasion, elephants in the circus get loose and run amok, killing their trainer and others who may be nearby. Of course, humans blame the elephant who dared to rebel this way instead of placidly accepting his or her treatment.

But other issues - hunting, poaching, extinction, factory farming, climate change – clearly affect and harm both people and animals at the same time. This book begins in Chapter 1 with a description of the utilitarian way mankind treats wildlife, to man's supposed benefit but the animals' detriment. The second chapter describes how human beings are harmed directly or indirectly by the way wildlife is treated and how many people would benefit if that were changed. The third and fourth chapters deal with marine wildlife: marine mammals, fish, corals, shellfish, etc. Chapter 5 goes into climate change in great detail because of its horrendous potential effects on human and animal life, and Chapter 6 describes the civilization and life-ending possibilities associated with nuclear war. Chapter 7 lists a few ways that we can start to turn things around, to the benefit of all.

In many cases the way animals are treated is sanctioned by powerful interests and even by governments, and trying to change minds and fight entrenched interests is a difficult and long term process. Animal welfare and animal rights organizations have been carrying on this battle for decades, and have made slow but steady progress. Cruelty to animals, for instance, was a minor issue in most U.S. states and generally didn't cover wildlife, but recognition of it as a moral issue, and perhaps the recognition of the link between animal abuse and violence to humans, has resulted in increased appreciation of the need to take animal cruelty more seriously. However, there's still a long way to go; many other countries and some religious faiths still place little consideration on non-human life.

Critics might say we have much more to worry about than our treatment of animals. People are quite understandably concerned about the economy and their jobs, retirement, the next political election, religion, immigration, healthcare, crime, and terrorism, yet people don't often see that how we treat the natural world, the air and water, the land and the forests, and generally speaking the planet and animals, has major long term consequences for people. Biologists have much to say about the balance of nature; damage one part of it and consequences can ensue in unexpected ways. Many species are connected, sometimes in surprising ways, from the apex predator to the lowest invertebrate. Removing a keystone species can have damaging repercussions cascading down an entire chain. Animals, flora, and invertebrates are intimately involved in maintaining the health of the environment and ultimately the health of the entire planet.

Connections between species and the complex role they play in nature are present all over the world. Some of these connections are apparent; some are more obscure. Excessive shrimp harvesting off the coast of Australia caused an explosion of the Crown of Thorns starfish that destroyed many of the coral that build and maintain the reefs that protect the Australian coast from storm surges. Wholesale destruction of coyotes causes population surges in small rodents that are the main carriers of ticks that carry a number of diseases. Birds and bats, injured by pesticides, are major predators of mosquitoes that carry serious and sometimes fatal diseases such as malaria, Zika, dengue, and chikungunya. There are many other examples where damage to one species inadvertently damages others.

The Middle Ages were racked by infectious diseases such as the "black death," the bubonic and pneumonic bacterial plagues that sprang up periodically over millennia and devastated human populations (another type of plague, smallpox, is viral in nature). Superstitious people decided that cats, the small, beloved species we welcome into our households to live beside us, were evil beings, perhaps minions of the devil and his human representatives on earth. European society began to eradicate cats until the population was drastically reduced. The consequence was an enormous rise in the rodent population that carried fleas that spread the deadly diseases. Reducing the cat population contributed to the deaths and sufferings of millions of humans; by some account a third of the population of Europe was wiped out. Cats were the allies that could have saved countless lives, yet we destroyed our very protectors because someone, likely due to misguided religious influences, decided they were evil.

Could it be that modern day humanity is making similar mistakes, that animals we are decimating now are playing beneficial roles that we don't yet understand and appreciate? Will we destroy the very saviors we should be supporting and thanking?

These lessons of history, ecology, and the enormous benefit of biodiversity appear to be lost to many people these days. In the United States, there is a war on wildlife, particularly the predators, the wolves, bears, coyotes, cougars, foxes, bobcats, alligators, and sharks. One of the war's greatest victims is the humble coyote. As mentioned earlier, a major prey of coyotes (and foxes) are small rodents that are the main carriers of ticks. Destroying coyote populations can result in higher tick populations, ticks that carry Lyme disease, bartonella, anaplasma, babesia, and several other diseases that are causing great suffering to many humans, diseases that are expanding in scope every year. Yet nearly every state in the nation encourages the virtually unlimited killing of coyotes.

"Whaling," the industry that has destroyed these magnificent ocean creatures for centuries and has thankfully been stopped in recent years by most countries, is still carried on by several rogue nations under one loophole or another in the IWC (International Whaling Commission) mandates. One of these nations even wants to kill whales in whale sanctuaries, supposedly inviolate areas where whales can breed, be protected, and increase their population. It has been found that whaling has inadvertently resulted in the increased activity of one of the sea's main predators, the orca, and the orca in turn has damaged the seal, sea lion, and sea otter populations. Sea otters in turn are major predators of sea urchins that feed on and destroy kelp and other seaweed where much of marine life would normally be born. As a result of this cascade, whaling can damage the fish population and fishing industry that so many humans depend on for seafood protein.

It seems that humans are predisposed to disliking "bugs," everything from beetles to flies to spiders to worms to bacteria. Collectively these invertebrates number in the many hundreds of trillions and we share the earth with them. Some people would like to see the bugs all gone, thinking the earth would be a nicer place. Yet science shows that humanity would probably be destroyed in a surprisingly short time without them. The complex role they play in the earth's ecology is sometimes obvious, but sometimes not. Worms we all know are the main force creating topsoil and contributing to the decay of waste matter, aiding in the growth of food crops that all life depends on. Bacteria in our intestines are essential for the digestion of our food. In a less obvious benefit, invertebrates prevent the excessive growth of fungi and other plants that could devastate human and animal life if allowed to proliferate. They are also the reason why we aren't destroyed by our own organic waste. These humble bugs and their invertebrate kin *literally keep us alive.*

Climate change is a long-term issue that will have some effect on those alive today but will have a much bigger effect on their children, grandchildren, and future progeny. Humans are in the process of altering the natural processes of the entire planet in ways that will become worse and worse as time goes on, unless something changes. The political debate on climate change rages on, not by scientists who concluded its reality years ago, but by factions that fear that any efforts to impact climate change will interfere with their short-term selfish economic and political interests. Animals and people could both suffer the catastrophic consequences of climate change, and helping animals by preventing this environmental devastation would benefit people enormously.

CHAPTER 1
Helping Animals

SPORT HUNTING

Different forms of life in different aspects of existence make up the teeming denizens of this earth of ours. And, no matter whether they belong to the higher group as human beings or to the lower group, the animals, all beings primarily seek peace, comfort, and security. Life is as dear to a mute creature as it is to man. Just as one wants happiness and fears pain, just as one wants to live and not to die, so do other creatures.

- XIV Dalai Lama.

Hunting would seem to be a simple and clear issue when it comes to helping animals; not wounding, killing, mutilating, or harming them is surely in the individual animal's best interest. There are many ways that cutting back on hunting and its related issues benefits both people and animals, issues such as deer-car collisions,

Lyme disease, canned hunts, poaching, extinction, the bushmeat trade, predator killing contests, and proscribed (intentional) burns. There can be much cruelty involved in these animal-killing practices, something which is largely hidden from the public. Neither hunters nor wildlife agencies ever seem to mention the suffering involved with the "sport." Events like the killing of the beloved and famous lion Cecil who took many hours to die wounded by an arrow and finally killed with another arrow raise public awareness for a short while, but this sympathy and awareness doesn't seem to transfer to other animals, even to other lions. Even the poaching (read - illegal killing) of magnificent elephants for their tusks to the tune of 96 a day, more than thirty thousand a year, doesn't seem to raise an uproar, or the serious possibility that the majestic tiger, surely one of the world's most beautiful animals, may be hunted and poisoned to extinction in the wild in a few years.

In the hundreds of articles I've read on the subject of hunting, I have hardly ever come across one that mentions the suffering that animal "targets" may experience (in fact, the only one I did come across actually argues that animals shot by crossbow or bow and arrow probably don't feel much pain, no more than a human working out at a gym [1]. The article was written by a hunter). On the contrary, most articles are glowing descriptions of hunting and killing of deer, pheasants, ducks, squirrels, rabbits, bears, coyotes, wolves, etc. where the enthusiastic writer can't say enough about the experience. Words commonly used to describe hunting include "harvesting" as if the living animal were a stalk of wheat, "culling" meaning to reduce populations, and "taking" as a euphemism for killing. Pictures appear in magazines and newspapers routinely of men, women, and children posing proudly next to the body of a bear, a deer, an elk or moose, or others, such as lions or leopards or cheetahs in Africa. One wonders if taking the life of a living being ever crosses these people's minds, or if the suffering their "targets" may have experienced gives any pause for thought. In a world where violence is on the rise everywhere and compassion is in short supply, the view of a smiling 9 or 10 year old child holding a rifle as big as the child, sitting next to a beautiful dead animal, brings sadness and anger to anyone who admires and appreciates wildlife. One can't help wonder what lesson this casual killing of a living being is teaching the child.

I was first exposed to the cruelty that can take place in hunting when I was about 10 years old. One of my father's friends who happened to be a hunter came to our house one evening. He proudly described his latest hunting experience, not realizing that my father wasn't likely to be impressed. He was deer hunting with either a shotgun or some high powered rifle. He said that when he shot the deer, somehow he shot off half of both front legs. The deer ran off as best he could stumbling on the stumps of his legs, trailing blood that made it easy to follow. Apparently the deer managed to run a long distance before succumbing to pain, blood loss, and shock. The hunter seemed rather annoyed at the deer as he told the story, that he had such a long way to go before retrieving his prize. When this 10 year old kid (myself) asked the man if the deer hadn't been suffering, he looked at me with a blank stare as if the question were either stupid or irrelevant.

My next experience was as a teenager living in Rio De Janeiro, Brazil. We had a guest speaker one day at the high school. He was a man who made a living hunting "tigres" as the Brazilian word for jaguars is called. He was a rather different kind of hunter, though. He hunted these jaguars, 200 pound ferocious cats that could easily kill a man, with a spear, believing that a true hunter had to give the hunted animal as equal a chance of survival and defense as possible. Many of the jaguars he killed were attacking the farm animals of local ranchers and occasionally the ranchers themselves, so at least there was some justification for his hunting. It wasn't *sport hunting* as is so much the case with hunters today, few of whom are subsistence hunters who live off their kills. The Brazilian jaguar hunter was saving lives by his hunts, and he was so in tune with nature that he insisted on giving the big predators a defensive chance. I didn't like it, his killing these magnificent animals, but I could understand its necessity. Contrast that with the American practice of "canned hunts" where so-called sportsmen can shoot a captured animal so tame they may have been feeding it moments before, while others literally shoot predators like bears while still confined in their cages. I'd like to think that most "real" hunters deplore the whole notion and practice of canned hunts.

Animals Killed by Hunting.

The number of animals killed by hunters in the United States is staggering. In data from 1975, the numbers for major species killed were: 94 million game birds (quail, doves, pheasants, turkeys), 32 million squirrels, 27 million rabbits, 21 million waterfowl (ducks, geese, swans), 2.6 million deer, 102 thousand elk, 84 thousand antelope, 67 thousand moose, 55 thousand caribou, and 24 thousand bears. Other kills include wolves, foxes, coyotes, cougars, bobcats, and boars [2]. By 2011 the number of deer killed had risen to 6 million and the estimated total number of animals killed in the U.S. to 200 million per year. These numbers don't reflect the collateral deaths such as bear cubs or coyote and wolf pups that die from starvation or predation when their mothers are killed. They also don't include the animals wounded but not killed, wounded victims that may die after lingering in pain from infection or starvation after days of suffering. In some studies, as many as half of a deer herd shot by bow and arrow were not "recovered" by the hunter [3]. A South Dakota Game Department biologist estimates that 3 million wounded ducks are unretrieved every year, and that's in South Dakota alone [4]. The Michigan Journal of Wildlife Management estimates that as many as 35% of deer rifle-shot by hunters escape wounded, and even higher numbers from bow hunting [5].

Hunters like to claim that they prevent animals such as deer from starving by reducing the population, and therefore serve as a balance of nature. It will be described later how hunting as mandated by wildlife agencies and practiced by hunters paradoxically doesn't reduce populations and is ecologically unsound. Populations aren't reduced because the mandated practice of killing mostly bucks while leaving females alone results in increased average births per doe the following year. Hunting is ecologically unsound because hunters take out the biggest, strongest, and fittest while natural predators would take out mostly the small, old, sick, and weak, leaving the fittest to breed and maintain a strong gene pool. Nature may not always be pretty, but if left alone, it's always smart.

The killing of top predators like bears, wolves, mountain lions, and coyotes, is vehemently encouraged by a proportion of the nation's private and corporate agriculture: the livestock industry, and other private interests. The federal and state governments through their wildlife agencies are only too happy to oblige. The pro-hunting bias of wildlife officials and game commissioners stems from their being financed by sales of hunting, fishing, and trapping licenses and matching funds from federal excise taxes on guns and ammunition, and most of these officials are present or former hunters [3]. Non-hunters aren't usually welcomed as members of wildlife agencies, even those agencies whose stated purpose is to protect wildlife.

Safari Club International has established levels of achievement for killing ever increasing numbers of animals of multiple species [6]. The Bears of the World Grand Slam requires killing an Alaskan Brown bear, a grizzly bear, a Eurasian Brown bear, and a polar bear, never mind that the last is an endangered species. The Diamond level of Achievement requires members to kill at least 322 animals, and the coveted Bronz Buffalo was awarded to a hunter who killed at least one of each of 369 species. The irony is that Safari Club is listed as a 501 (c) (3) organization organized for exclusively charitable and educational purposes.

In many states, government wildlife agencies have initiated school programs to indoctrinate children at an early age into hunting, trapping, and fishing, with classroom presentations and contests for children who write about the "need" for game management [2]. The U.S. Fish and Wildlife Service (USFWS), a federal government agency, also has programs to influence children toward hunting and away from empathy toward wildlife [2]. At least twenty-nine states have sponsored special children's recruitment hunts on public land, and thirty- eight states have used hunter training as part of junior and senior high school curricula [7]. Children in these early ages, especially the elementary school years, 5th to 8th grades, are impressionable and easily indoctrinated when presented with only one side of an issue. If humane education were given equal time, children would at least have a balanced view and could make up their minds how to view wildlife. These government agencies have the major advantage of appearing "official" which adds unequal power to their propaganda. Humane education is most often carried out by a few private volunteers, if it is carried out at all.

Teaching children to enjoy killing animals through hunting may have unexpected and serious long-term consequences for humans in some rare cases, part of the well-known connection between animal cruelty and human violence, though government agencies vehemently deny that wounding or killing animals is cruel, turning a convenient blind eye to any suffering involved. The lack of compassion for animals has been tied to a number of attacks and deaths of people from individuals apparently desensitized to harming other living beings through hunting [2]. What is the lesson being taught to school children when the "officials" encourage them to hunt and kill nature's living beings? At a time when society would benefit by encouraging kindness, empathy, and prosocial values in our impressionable young people, we teach them to condone and possibly participate in killing instead. Is this really what we want?

WILDLIFE KILLING CONTESTS

It is Man who has upset the balance of nature, killing off certain predators, moving animals to where they don't belong and have no natural enemies, all this for his own misconceived purposes. Man considers a pest any animal that doesn't put money into his pocket. So war is launched upon the animals with disease, viruses, chemicals, and guns, all to remedy Man's own blunders.

- Dr. Douglas Latto

For sheer killing in the name of fun, probably nothing can beat the ever-more-popular hunting *contests*, particularly predator killing contests. These are often sponsored by sportsmen's clubs but sometimes by companies selling hunting gear, weapons, and other outdoor equipment. All across the U.S., there are organized hunting contests to see who can kill the most rabbits, squirrels, turkeys, and predators like foxes, bobcats, and coyotes. These contests have exploded across the country in recent years [8]. Prizes are given out for the largest animal killed, the smallest, the most individual animals, the most animals by weight. Towns and cities in many states have killing contests in the name of charity or for fund-raising, with squirrels, rabbits, or other small animals as the victims.

Coyote killing contests are particularly common, held sometimes once a year and sometimes more often in many different States. The website coyotecontest.com lists past and upcoming killing contests, which take place mostly in the West and Midwest but also in New York, Ohio, and North Carolina. In each contest, there are no limits on the number of animals that can be killed in an established time period. As pointed out in several websites, the federal and most state governments allow coyote killing at any time and in any number, on the rationale that coyotes might kill livestock.

Having succeeded in destroying most wolves, and continuing to allow hunters to shoot them almost all year around, federal and state governments have set their sights on destroying as many coyotes as possible. "No other wild animal in American history has suffered the kind of deliberate, and casual, persecution we have rained down on coyotes. … There is something perverse in the government, and society, marking a species for death, setting it outside the bounds of even our wildlife protection laws" [9].

According to www.predatordown.com, Utah even has a Youth Coyote Hunt, where children get to participate in the fun. One parent describes the success of his 11 year old daughter and 9 year old son in killing coyotes. Another, the creator of the Coyote Craze website, proudly describes how two of his sons made their first coyote kills at age 5. The West Texas Big Bobcat Contest awards $76,000 in prize money for the most coyotes, bobcats, and foxes killed either by single hunters or in teams. Arizona has a "Santa Slay" contest around each Christmas to see who can kill the most coyotes during the yuletide season.

Endangered wolves look very similar in appearance to coyotes and are sometimes shot by accident in these predator killing initiatives. In a side by side comparison, it is hard for even outdoorsmen to accurately tell them apart. Bear killing contests also exist in many states, though not with the magnitude of the coyote contests as the bear population is much smaller. Other contests occur frequently to kill the most squirrels and rabbits [10], contests which are in *addition to* the normal small game hunting season. And, of course, there are "big game" hunting competitions in 48 states [11], mostly deer but also elk, moose, caribou, bighorn sheep, mountain goat, and antelope. Deer are by far the favorite target of hunters, and contests are designed to award prizes for the biggest, strongest, and best specimens, particularly bucks.

Predators are not the only subjects of these contests. Idaho in past years held a "bunny bash" festival where people of any age but especially young people were encouraged to beat captured rabbits to death with tire irons, baseball bats, and clubs, even playing bunny baseball where youngsters competed to see who could hit rabbits the farthest [2]. Contests have also been held to see who could kill the most prairie dogs. All done in the name of fun!

Then there are the pigeon shooting "festivals." Not content with "trap" shooting clay targets, sportsmen vie to shoot live birds instead, and in some states, regular contests are held to see who can kill the most pigeons, which are much slower than clay targets launched by machine and easier to hit.

As has been said earlier, what kind of lesson can we be giving young people, some as young as 7, 8, or 9 years old, when we teach and encourage them to beat as inoffensive an animal as a rabbit with baseball bats or tire irons, or to see how far they can hit the living, suffering creature in a game of rabbit baseball? Is this really the upbringing of choice for our future adults, leaders, husbands and wives, fathers and mothers?

PHEASANT STOCKING

Wild animals never kill for sport. Man is the only one to whom the torture and death of his fellow creatures is amusing in itself.

- James Froude.

In case there are not enough targets to shoot, many states have programs in which they raise tame birds on game farms then release them onto specific hunting fields where hunters wait to shoot them.

This is most commonly done with pheasants but sometimes with quail. At last count, the states of CO, CT, IA, IN, MA, ME, MI, MT, NH, NJ, NY, OH, PA, RI, WI, and WY all had pheasant "stocking" programs for the benefit of hunters, since in many cases the populations of wild pheasants are too small for large scale hunting programs. Pheasants are raised from chicks on farms until they are 8-12 weeks old, then released at the start of pheasant hunting season. A Youth Weekend often takes place before the regular season to give children ages 10 and up the chance to make their kills without competition from the adults. One common practice is to spin the birds around by their feet to make them dizzy just before release so they don't fly away until flushed out [12].

According to pheasantsforever.com, the tame birds have no knowledge of finding food, escaping predators, and otherwise surviving in the wild. If they survive the shoot, only 60% last the first week, 25% the first month, and 5-10% for longer periods, particularly in cold weather states [13]. Most deaths of these tame pheasants come from predators, even more than are shot by hunters. The low survival rates mean that the stocking programs have to go on year after year to provide the desired number of targets.

A blatant example of pheasant stocking (and canned hunts) came when former V.P. Dick Cheney and nine of his friends were reported to have killed over 400 tame pheasants and an unknown number of ducks at a shooting ranch in Pennsylvania [14]. The V.P. was credited with shooting 70 of the birds by himself. He is reported to have flown to the event in Air Force Two. Perhaps there was other business in the area to justify the use of the big jet plane.

WILDLIFE SERVICES
(FORMERLY ANIMAL DAMAGE CONTROL)

As long as man continues to be the ruthless destroyer of lower living beings, he will never know peace. For as long as men massacre animals, they will kill each other.

- *P*ythagoras of Samos, 500 B.C

It was first of all necessary to civilize man in relation to his fellow men. That task is already well advanced and makes progress daily. But it is also necessary to civilize man in relation to nature. There everything remains to be done.

- Victor Hugo.

If there were ever an agency incorrectly named, it is Wildlife Services, as if it altruistically existed for the *benefit* of wildlife. Once it becomes known and understood what this agency is, the more apt name would be "Wildlife Killing On Demand Services." It was formerly known as Animal Damage Control until 1997, and its supposed purpose was to answer complaints from the public about wildlife problems, particularly complaints from ranchers about possible livestock predation. A brief look at the list of animals killed every year might cause some degree of skepticism; in 2014, among the 500 species or subspecies of animals killed using taxpayer dollars and in the supposed public interest were turtles, toads, frogs, ducks, mice, iguanas, endangered bald and golden eagles, parakeets, doves, cardinals, otters, squirrels, bluebirds, and chipmunks (I guess there must be killer chipmunks out there), besides 470 other types of animals. Along the way, many household pets are killed accidently and sometimes not accidently, particularly dogs.

Wildlife Services is a division of the Department of Agriculture.

It was formerly a division of the Department of the Interior, which would seem more appropriate since wildlife are part of Nature's heritage, but ranchers were dissatisfied that the DOI wasn't forceful enough in killing predators, so they lobbied Congress to make the change. Its mission statement is "to provide Federal leadership and expertise to resolve wildlife conflicts to allow people and wildlife to coexist," and as part of that it has programs in rabies eradication by oral vaccine distribution, elimination of harmful non-native invasive wildlife that may harm native species, and resolving bird issues at airports to prevent airplane/bird collisions. However, as has been noted in many investigations from the GAO (Government Accountability Office), investigative journalists, and former WS employees, the agency has morphed into a killing agency. Over the past 18 years, it has killed 2.8 million animals a year, with 4.3 million in 2013 [15] and as many as 5 million in other years [16], a total of more than 50 million animals [17]. All this is at the mere request of the members of the public when they claim these animals are pests or nuisances.

Killing methods include shooting, leghold traps, neck, leg, and body snares, aerial shooting from planes and helicopters, denning (killing baby animals in their dens), and widespread use of poison. Poisons in particular have a long, dark history. Poisons, including thallium, strychnine, cyanide, and compound 1080 were tested extensively but were banned by President Nixon in 1972. Sodium cyanide was reauthorized by Pres. Ford in 1975. President Reagan and his Interior Secretary James Watt then revoked the general ban on poisons in 1981. Today, cyanide is used in M44 cartridges placed in the ground and scented with food to attract predators. When they bite on the cartridge, a lethal dose of cyanide is injected into their mouth. The cartridges are indiscriminate, killing a range of animals including pet dogs and endangered eagles in addition to their target coyotes. Compound 1080 is one of the deadliest poisons known, lethal to humans in sizes comparable to a few grains of rice. It's placed in collars around the necks of sheep or goats so that coyotes or other predators that may attack them receive a lethal dose and die after 6-8 hours of suffering. Strychnine has been used to kill prairie dogs by distributing it in bait food and causes intense stomach pain until death comes as a relief.

Leghold traps, snares, and pain-inducing poisons would surely come under cruelty statutes if they weren't carried out by a government agency or, in the case of traps, encouraged as an outdoor "sportsman's activity" by state wildlife agencies. However, *illegal* cruelty is no stranger to Wildlife Services either. In 2012, a Wildlife Services agent posted pictures on line of his dogs attacking coyotes caught in leghold traps. When the outrage came out, a former WS agent whistleblower stated that it was common practice for agents to let their dogs maul and kill trapped coyotes, and everyone in the agency knew about it. He also described how his supervisor accompanied him one day as he approached several coyotes caught in snare traps and laughed as the dogs ripped the nearly helpless animals apart. WS agents have also let their dogs maul and tear apart newborn mountain lion kittens while hunting their mothers [18].

The same whistleblower WS agent described that on many occasions, endangered and federally protected golden and bald eagles were caught in leghold traps, snares, or killed by the M-44 explosive cyanide cartridges, and how WS management instructed field agents to bury the eagles and "keep their mouths shut" [19-22]. Another former agent described how his supervisor forced him to participate in testing the cyanide death technique on dogs from a local shelter, forcing it into their mouths until near death, bringing them back with an antidote (amyl nitrate), then forcing more cyanide into the mouths of the same dogs to repeat the process [19, 22].

If these were isolated incidents, it would be bad enough, but animal cruelty against coyotes appears to be common practice in

Wildlife Services [19, 22-23]. In the aerial shooting of coyotes, many are wounded instead of killed. "Some of the gunners are real good and kill coyotes every time. And other ones wound more than they kill. Who wants to see an animal get crippled and run around with its leg blown off? I saw that a lot," said Gary Strader, former WS agent [23].

M-44 cyanide gas cartridges have killed many non-offending animals, including over 250 dogs [24]. When traps and snares are added, the toll to dogs exceeds 1100 [18]. On average, eight dogs a month were killed by mistake, a figure a former agency employee believes is low. "We were actually told not to report dogs we killed because it would have a detrimental effect on us getting funded," stated former agent Rex Shaddox [18]. Many of the dogs killed were family pets, since poisons or traps were sometimes set close to residential areas.

Wildlife Services has been called an agency "out of control," seemingly answerable to no one, including members of Congress [24]. The lack of transparency is carefully guarded by the agency, and leads to some absurd consequences. For example, some of the animals killed by the agency are the focus of conservation and restoration efforts [18]. "The irony is state governments and the federal government are spending millions of dollars to preserve species and then Wildlife Services are out there killing the same animals" [25]. Another hidden fact is that people have been injured by the use of the cyanide M-44 cartridges, and at least 10 people killed and many others injured by plane crashes during aerial shooting of predators by WS agents [18].

Wildlife Services carries out its programs with little regard to sound ecological principles, especially the very important and often under-rated "balance of nature." Every species, including coyotes, have a role to play in that balance. "In the Northeast, for example, the elimination of red wolves led to a proliferation of coyotes," which normally reduced the population of foxes, which prey on small rodents like field mice, which are the main carriers of ticks that carry Lyme disease bacteria as well as several viruses and parasites [15]. The explosion of these small rodent populations has greatly increased the occurrence of these bacteria and viruses that can cause life-threatening and lifetime-debilitating illness for human beings. The elimination of wolves and bears results in higher populations of deer, moose, and elk that destroy young trees that many other animals depend on for habitat. In Yellowstone National Park, the reintroduction of wolves to counter a dense overpopulation of elk that had been overgrazing on aspen, willows, and cottonwoods brought back many other species such as beavers, songbirds, and other wildlife. These are just a few examples of the "balance of nature."

To make matters even worse, it isn't clear that the nonselective killing of predators is effective. A study of mule deer populations in an area where predators had been killed off compared to an area where they hadn't, showed no difference in the deer populations the following year. Presentation of such findings to WS personnel had a tendency to lower morale, as if their predator killing wasn't of much if any benefit. But killing things to solve problems seems to have broad appeal as a wildlife management practice, providing quick visual feedback to the manager and the farmer, even if it isn't very effective in the long run. "The simplicity of predator control has broad appeal. The complexity of the problem is far greater," said Tony Wasley, a Nevada Wildlife Department biologist [23]. USDA statistics show that most livestock problems and losses result from weather, disease, illness, and birth problems rather than predation [26]. Nonlethal methods of protecting livestock, such as guard dogs and fences, were as effective in cases that were studied as were lethal methods, and probably a lot cheaper to the taxpayer [27].

And just as deer adjust to population decreases by increasing the average number of fawns born per doe, so do coyote populations rebound by increasing the number of pups born in a litter [23, 27], so the population doesn't really decrease despite all that killing. As said earlier, nature isn't always pretty but it is always smart.

POACHING.

I have come to realize that the dominion which we have been given over the animals is a serious responsibility. It does not mean that we are justified in exploiting the animals, and taking their lives and causing them pain and terror, and treating them as chattels for our profit or sport or amusement. The animals are our younger brothers and sisters on the same ladder of evolution as we, and we are responsible for helping them up the ladder instead of retarding their development by cruel and callous and greedy exploitation.

- Air Chief Marshal Lord Dowding

Whatever one thinks of hunting, it is legal and sanctioned by the governments around the world. Wildlife managers and agencies supposedly examine the data on populations and expected future wildlife trends and set quotas for the number of animals that can be legally "harvested." In addition to legal hunting, there is an ongoing worldwide Wildlife Trafficking sanctioned and supported by governments. Organizations such as CITES (Convention on International Trade in Endangered Species) and IWC (International Whaling Commission) have been set up to ensure that the very large trade in various species doesn't represent over-exploitation and cause a danger of extinction.

Poaching is illegal hunting (killing), illegal wildlife trafficking, not sanctioned, not planned, and not controlled by CITES or any other legal organization (poaching of endangered trees and plants also takes place). In theory, sport hunting would not cause species endangerment or extinction, but poaching is the second biggest threat of extinction after habitat loss. Hunting is carried out by millions of individuals in an activity they call "sport" and hunters for the most part (with glaring exceptions such as canned hunting) have respect for nature and even for the animals they kill. In many cases today, poaching is controlled by organized crime associations and even terrorist organizations [28-33]. These organizations will quite likely continue until the species being killed is either extinct or so reduced in numbers as to remove the profit; then they will move on to another species to start the process over again. In effect, poaching is one of the key drivers toward endangered species and extinction. It is estimated that illegal wildlife trafficking is a $6-20 billion/year business [33-36]. The close connection with the illegal wildlife trade and organized crime and terrorism organizations makes poaching a U.S. national security issue [34].

A great deal of the wildlife trade is carried out to fuel Asian markets. The list of Asian countries participating as either markets or suppliers of poached wildlife includes China, Taiwan, Japan, Viet Nam, South Korea, Hong Kong, Burma, Laos, Cambodia, Malaysia, India, and Russia [36]. The U.S. is a large consumer of both legal and illegal wildlife [35, 37], even though U.S. endangered species laws should minimize or prevent the trade entirely. The U.S. is the 2nd largest consumer of poached ivory and other wildlife products and a conduit of products going to China [34-35]. Ivory, rhino horn, tiger body parts are seen either as status symbols or as components of traditional medicines to cure headaches and hangovers in Asian countries [28, 36, 38-41]. Rhino horns are even made into drinking cups as a status symbol for the wealthy [34].

Another example of wildlife devastation due to poaching comes from the bushmeat trade. The trade is built upon poachers killing large numbers of chimpanzees, gorillas, bonobos, pangolin, elephants, crocodiles, and antelopes [42-43]. Killing methods include wire snares, poisoning, firearms, and hunting dogs. Much of the meat from the poached animals is sold in local logging camps, but some of it is smuggled into the United States and parts of Europe. The problem has been exacerbated in recent years by road construction through forests for the logging and mining industries. Eating bushmeat has become a status symbol for wealthy Africans, which has also increased the demand. Gorilla meat in particular is considered prestigious for wealthy customers [44]. Gorillas are facing insurmountable odds from habitat destruction, bushmeat poaching, diseases such as ebola, and the demand for gorilla trophies, and this wonderful species, so close in many ways to humans, could be extinct in a few decades [45].

There is not much interest inside African governments in fighting against poaching; either it is too difficult to stop or represents lucrative kickbacks to local officials. The same international crime organizations that smuggle weapons and drugs are involved in poaching and trafficking wildlife along with money laundering and racketeering [31]. In some areas, park rangers try to protect wildlife but are outnumbered by the poachers, who have better equipment and more powerful weapons. Over 1000 park rangers have been killed in the decade before 2015, an average of 2 rangers killed every week [31-32]. (More on this danger to humans will be described later).

Elephants.

The iconic species endangered by poaching is the African Elephant, killed almost entirely for their tusks (ivory) but sometimes as bushmeat. Between 35,000 and 40,000 elephants are killed each year, around 96 a day or one every 15 minutes [30, 37, 41, 46]. In Kenya, the elephant population has gone from 175,000 initially to 35,000 as of 2015, an 80% decline, reportedly due in part to the terrorist group Al Shabaab [30]. In one horrible event, an army of poachers in Bouba Ndjida National Park in Cameroon killed 650 elephants, the entire herd including the babies, using AK47s and Rocket Propelled Grenades. The babies were tortured to draw in the adults trying to protect them [47].

Mainland China is the world's largest consumer of ivory, considered a wealth and status symbol. Much of the trade has been passing into China through Hong Kong, which has taken steps in recent years to curtail the traffic [32, 48-49]. As mentioned earlier, America is the 2nd largest consumer of ivory after China, and therefore the 2nd largest cause of elephant death and possible extinction [35, 37]. Both China and the U.S. have announced plans to fight the ivory trafficking problem, alongside a campaign to encourage people not to buy wildlife products [30-31, 35, 37, 49, 50]. The proposed ban on ivory in the U.S. is opposed by groups like the NRA (National Rifle Association) which also opposes state bans on the trade [30]. In the past, the ivory trade has continued in spite of proposed bans due to loopholes in the restrictions, particularly allowing the sale of "old" ivory products obtained before the elephant was listed as endangered, or if the elephant died of natural causes. The loopholes have been wide enough to result in continuing the trade almost unabated. It will be up to the U.S., China, and others to see that new rules and restrictions are seriously enforced and loopholes aren't exploited. A very promising step was taken in late 2016 when China announced it would ban the ivory trade by the end of 2017, cutting off probably the poachers' biggest markets [51]. It is very important that other Asian nations don't substitute for China's ending the ivory trade, and that China follows through on its announcement without loopholes or exceptions.

Rhinos

Even more critically endangered than elephants are rhinos and tigers. Rhinos are killed strictly for their horns. Many Asian cultures have the notion that powdered rhino horn is a cure for various diseases [39, 41], including cancer, hangovers, fever, and impotence. Viet Nam is the largest importer of rhino horn [29, 39]. The Western Black Rhino has already been hunted to extinction [38] and the other 5 subspecies are endangered or critically endangered. Several remaining rhino populations are down nearly 98% [41]. Nearly 1200 rhinos were poached in 2014; this is a significant fraction of the remaining population [29]. Rhino horn can bring up to $60,000 a pound on the Asian market [38]. Rhino horn is the same material as human fingernails which have equal medicinal non-value.

Tigers

There may be only 3200 tigers left in the wild as of 2015 [28, 36, 40] where there were originally 100,000, representing a 97% decline. It appears to be inevitable that tigers will be extinct in the wild within the next generation. The driving force is the Asian market where the culture considers tiger parts to be useful for "traditional medicines." China, Taiwan, Japan, South Korea, U.S., and UK are involved in the tiger trade [36]. China has already eradicated its own wild tiger population so is constantly seeking new suppliers. Even though China is a member of CITES, it largely ignores CITES rules [36]. Tiger poaching is also extremely inhumane, involving poisons, leghold traps, snares, firearms, or electrocution [33, 52]. One of the biggest tiger markets is in Japan, while Hong Kong is the main supplier for mainland China. Russia has also become a key supplier for the Chinese market.

In addition to medicine, tiger meat is served at special functions, as a sign of status and wealth, and tiger "trinkets" like teeth and claws are sold in local stores. Parts from a single tiger can bring $50,000 on the black market, and tiger "farms" have arisen in China, Laos, and other parts of Southeast Asia [40], with reportedly more than 200 tiger farms in China alone. Just as in the lion farms in Africa, these magnificent great cats live in miserable conditions and don't bear much resemblance to their wild cousins. Recently, over 130 tigers were found on a tiger farm at a popular Buddhist temple in Thailand, together with the dead bodies of 40 tiger cubs found in a freezer [53]. Authorities confiscated the tigers after allegations of wildlife trafficking and abuse came to light, though the monks denied any wrongdoing.

Gorillas

The world's only wild gorilla populations live in one area of Africa, in national parks in Uganda, Rwanda, and the Democratic Republic of Congo. These countries have laws protecting these populations but the laws are ineffectively enforced, saddled with apathetic judiciary systems, or subject to corruption. By some estimates there are less than 900 individuals remaining in critically endangered gorilla populations [41, 54] and gorillas are being destroyed faster than they can reproduce. Gorillas are dying from habitat loss, diseases, and poaching. Gorilla infants are sold for $40,000 each and gorilla parts serve as trophies or for medicines and magical charms [44-45]. Gorilla heads are favored trophies and gorilla paws serve as ashtrays.

Ebola affects all the great ape species. By some accounts, the disease has destroyed one third of the world's remaining gorillas and chimpanzees [55]. In 1994, an ebola outbreak in Gabon wiped out the world's second largest protected population of gorillas and chimpanzees. Ebola remains a major threat to the remaining great apes, but an even greater threat is the bushmeat trade as already described. Not only are gorilla and chimpanzee bushmeat [56] consumed locally, but gorilla meat has shown up in Paris, London, New York, and Miami, smuggled in suitcases through major airports [45]. The bushmeat trade worldwide is estimated to be worth $1 billion annually [57]. Often leg or arm-hold traps are used to capture great apes, extraordinarily cruel devices for any species.

More than 5 million tons of bushmeat are taken from the Congo Basin forests each year [57]. Chimpanzees, gorillas, bonobos, and other ape species are helpless against humans with high powered rifles, leghold traps, and poisons. It's hard not to see the destruction of the great apes for meat as a form of cannibalism, as they seem so close to human. Chimpanzees and gorillas have learned to "talk" in sign language. Koko the gorilla speaks over 1100 "words" in American Sign Language [58] and can carry on conversations with humans and with another gorilla. Koko and others like her, even from other species (like Alex the grey parrot), are proof of what many animal behaviorists refuse to concede, that animals can think, communicate, and express their thoughts if given the means and opportunity.

Sharks

Shark finning is another decimation of animals fueled by the Asian market and Asian culture. In shark finning, the main dorsal fin of a captured shark is cut off with knife or saw and the still living animal is thrown back into the sea to die a slow and likely painful death. The fins are used to make shark fin soup, considered a delicacy and status symbol in China and Japan [59-60] as well as a tradition. As many as 100 million sharks are killed each year, a death rate significantly greater than their reproductive rate, with the result that sharks are in danger of extinction [59]. Some shark populations have already decreased by 60-70% due to human shark killing.

Whatever one thinks of sharks, they play a valuable role in marine ecology, and removing such large numbers is likely to cause considerable damage to the ocean ecosystem [59]. Fewer sharks mean more rays which eat more scallops, clams, and bivalves which damages biodiversity in overt as well as subtle ways and damages human fisheries. "Sharks play a critical role in the ocean environment," said Pew's [Pew Environmental Group] global shark conservation manager, Jill Hepp. "Where shark populations are healthy, marine life thrives. But where they have been overfished, ecosystems fall out of balance" [60].

Some progress has been made in recent years. In early 2013, CITES listed several species of shark under their Appendix II as potentially endangered, and some countries are beginning to curtail or ban the practice of finning. In 2012, the Chinese government prohibited the serving of shark fin soup at official banquets and the United States has taken steps to conserve shark populations, requiring that the whole shark be brought to shore with the fins intact, significantly increasing the costs and logistical difficulties of shark fishing. Japan, however, continues to defy shark conservation and continues shark finning unabated [60].

Whales

Probably the most contentious poaching issue and one that has lasted many years is the issue of whaling, once carried out by many nations but at the moment largely restricted to Japan, Norway, and Iceland. Whaling has been regulated by the creation of the International Whaling Commission in 1946 [61], composed of 88 member nations somewhat like a miniature United Nations. In 1986, concerned by the large whaling kills and uncertain about the remaining populations, the IWC established a moratorium on whaling which still remains, though Japan, Norway, and Iceland continue to slaughter whales through loopholes in the regulations. Japan has been attempting to stack the IWC with new members that would rule in its favor to legitimize and expand its whaling operations. They have even announced plans to begin whaling in the Southern Ocean Whale *Sanctuary*, in spite of a ruling by the International Court of Justice to ban whaling in the sanctuary [62].

At least 50,000 whales have been killed since the moratorium was started [63]. The three whaling nations kill 2000 whales each year under the loopholes and continue to trade in whale products. Whales or their antecedents have existed in the seas for 50 million years. The process of killing these great creatures is also extremely cruel – exploding harpoons that could hit them anywhere, subjecting them to a slow and painful death. Whales are slow to reproduce and are always on the brink of becoming endangered; poaching along with ship strikes, ocean pollution, unsustainable fishing practices, and the ever-increasing effects of climate change all combine to threaten these highly intelligent beings that play a significant role in the health of the ocean ecosystem [64-65].

In the name of "tradition," inhabitants of the Faroe Islands, part of the nation of Denmark, slaughter hundreds of pilot whales and dolphins every year in a ritual they call the "grind." Ostensibly to carry out the slaughter for meat, it has been found recently that the whale meat is dangerously contaminated with heavy metals and organic pollutants including mercury and PCBs [66]. This contamination was found to contribute to birth defects, high blood pressure, arteriosclerosis, diabetes, and Parkinson's disease in residents of the islands. Similar health concerns apply to the dolphin slaughter traditionally carried out in parts of Japan, where 20,000 dolphins a year are slaughtered by Japanese villagers for meat and for sale to aquariums and "swim-with" programs [67-68]. The dolphin killing is fully supported by the Japanese government [68]. Helping these animals would clearly result in helping people.

Lions

The world may have been horrified and "up-in-arms" about the killing of Cecil the lion, but the world remains silent about the devastation of wild lions in Africa, where estimates are that only 20,000 remain in the wild, only 10% of their former population [43]. They are now extinct in the northern part of Africa. Habitat destruction, prey loss, the bushmeat trade, persecution by local farmers, and hunting are all contributing to the loss. Americans make up 60% of the lion hunting clientele. Demand for lion parts in Asia [30]: China, Laos, Viet Nam, is fueling the killing of lions, much of it from lion breeding farms much like the tiger farms in China. East Africa remains one of the last strongholds for wild lions, but in some areas such as Maasailand (Kenya), lions are being speared and poisoned at a rate that is likely to cause local extinction in a generation [69].

Bears

The bear taxon (group of species) is not endangered as a whole but specific species are, largely due to hunting, poaching, and habitat loss, but their greatest threat is in the Asian wildlife trade in bear parts: gall bladders, bear paws, and bear bile [70], particularly in South Korea and Japan. All the world's species of bears have experienced population declines because of this trade (except the giant panda). A gallbladder can sell for $10,000 on the black market, and bile is used in "traditional" Asian medicines as a magic cure-all. Bear paws are served as a delicacy at Japanese business banquets, and the *extraordinary cruelty* of this practice is exemplified by the act of cooking the paws over hot coals while still attached to the bear and while the bear is still alive [70].

Bear farms have been established all over China and other Asian countries (Korea, Laos, Viet Nam, Myanmar) to extract bile from bears kept in extremely cruel conditions similar to pig gestation crates or veal crates where the animal can barely move. They may remain in this condition for years. The practice still continues in these countries despite token laws prohibiting the practice. Bile is extracted from the caged animal from open, often infected wounds from tubes penetrating into their gall bladders. Both gallbladders and bile are sold in most major U.S. and Canadian cities which therefore help to keep these bear farms in business. The Bear Protection Act which would prohibit the trade in bear parts in the United States has been introduced multiple times and supported by animal welfare groups of all kinds. In spite of the known cruelty, this Act to protect bears is opposed by the NRA, Congressional Sportsmen's Foundation, Safari Club International, and nearly every group opposed to helping wildlife [71-72].

Attempts are also underway to remove Endangered Species status from Grizzly Bears so they could be shot outside of Yellowstone Park [73]. When similar federal protection was removed from the gray wolf, the wolf population was decimated with a loss of thousands in just a few years [73].

Investigations have uncovered large commercial organizations involved in poaching and smuggling bear gallbladders and bear paws *inside the U.S.* [71]. Significant poaching and smuggling have been uncovered in several states including Oregon and Virginia. Estimates have been made that 40,000 American black bears are illegally poached in the bear parts trade every year in North America [71]. Estimates are also that the bear parts and bile trade may be valued at $2 billion, even though there is no medical evidence that bear bile has any medicinal value, and even many Chinese doctors oppose its use.

CANNED HUNTS

Every particle of factual evidence supports the contention that the higher mammalian vertebrates experience pain sensations at least as acute as our own. To say that they feel less because they are lower animals is an absurdity; it can easily be shown that many of their senses are far more acute than ours.

- Richard Serjeant.

I sometimes wonder about the language of hunting. I question whether "game" or "sport" should be used to describe an activity in which one of the players has only his pure wildness to call on for defense, while the other has weapons of shocking power, lures and decoys, a hunting dog, telescopic vision, .. the ambush or blind, and a colored map of the migratory path of the prey. It is a game in which the stakes are life and death, but only for one of the players.

- Victor B. Scheffer

Canned hunts really aren't hunts at all. They're the shooting and killing of various species of animals penned in an enclosure that have no chance of escape and are often so tame they may be eating out of the shooter's hand moments before he kills them. The "hunts" are carried out in fenced-in areas ranging from a few acres to several thousand acres, but the central characteristic is that there is no escape for the animal, no chance to avoid their own killing or their own suffering. Canned hunts have been around for over 60 years, starting in Texas from an old cattle and antelope ranch. Many animals in canned hunting ranches were born and bred for the sole purpose of being killed, and many are so tame they have lost any fear of humans and are easy targets. The analogy with "shooting fish in a barrel" has been made many times. There are now over 1000 such ranches in the U.S. in 28 states.

The list of animals that are targets on these ranches range from the familiar deer, goats, and sheep, to the exotic predators. The variety of species worldwide includes deer, sheep, goats, antelopes, pigs, lions, tigers, leopards, blesbok and springbok, crocodiles, elk, oryx, zebra, rhinos, gazelle, bison, cougars, bears, and even elephants [74-78]. The animals may be former pets, obtained from wildlife auctions including exotic animal auctions featuring predators and endangered species, sold to ranches by dealers, or sold to ranches by zoos. Even some of the largest and most prestigious zoos sell excess animals either to middlemen dealers or to game ranches directly, animals that wind up in canned hunts. For zoos, animal sales bring in substantial funds and make way for breeding baby animals which in turn attract the public and increase attendance [74, 77, 79-80]. The hunts themselves are big business. Each animal killed may cost the shooter $5,000 – $20,000 and the more rare or exotic the species, the higher the price.

There are no Federal laws against the practice, even in the case of endangered species, and even though the Endangered Species Act should otherwise apply. The federal Animal Welfare Act also doesn't apply to canned hunt ranches. Ranch owners often have permits from the Federal Govt. for breeding and importing of endangered species. Monitoring of canned hunting ranches, if any, is left to the states. The USFWS can and does issue permits that allow customers of these ranches to kill endangered and threatened species. As long as customers are willing and able to pay, there are no bag limits on the number of kills, and customers don't need firearm experience or hunting licenses [77]. This virtually ensures many instances of wounding and suffering of target animals. In some cases, the animals are drugged [74] to make them sluggish and easier to kill, even for the poor marksmen.

An example of this has been described in the case of a Corsican Ram. A so-called hunter shot the ram in the flank with a 1st arrow, a 2nd arrow in the side as the ram ran to escape, a 3rd arrow in the side as he came up against a fence, a 4th arrow in the back, and 5th and 6th arrows in the side, until he fell over on the ground in great pain but still wouldn't die. After waiting some time, the customer shot the ram with a rifle, and began pulling the arrows out before the ram was even dead [74, 80].

Another example of the cruelty involved in canned hunting comes in the form of a pig, raised from birth as a favored and loved pet. Fred, a 1000 pound pig raised as a family member, pampered and treated with affection, was sold to a canned hunting ranch that *advertised him as a wild boar.* A man and his 11 year old son decided to "hunt him." The boy stalked and shot Fred 8 times with a handgun over a 3 hour period before Fred finally died [81-82]. None of the 8 shots was a kill shot, that could have been delivered at any time, and the multiple shots to his sides "would have caused extreme pain and anxiety for a prolonged period of time" (Dr. Y. Lee, 82). As a very smart animal always treated with kindness before, Fred must have wondered through his suffering what he had done wrong (and for you animal behaviorists who doubt that animals can think and feel in their own way, try living with a dog or cat or horse someday).

For dangerous animals, such as predators sold by zoos or circuses, exotic animals from wildlife auctions, or former "pets," the canned hunter can shoot them while still in their cages or pens [74, 79, 83]. In some cases, bears have been shot in their cages after state wildlife officials claimed that it would be too dangerous to let them out [79]. A video filmed in South Africa shows a man in the back of a pickup truck shooting a tame, almost playful lioness with a bow and arrow as the truck stalks her in her pen, two employees with rifles at the ready in case the lion should dare to fight back [84].

Perhaps the least known and biggest canned hunting operation worldwide involves the killing of lions in Africa, as well as tigers and leopards to a somewhat lesser extent. The outrage that the public felt over the killing of Cecil the lion doesn't appear to extend to the thousands of equally deserving and innocent lions killed every year in these killing operations. Lions are bred on wildlife farms in Africa for the sole purpose of being shot [76, 83, 85]. The driving forces are jointly the trophy industry and the Asian market for body parts, especially bones which are used in Asian bone wine. There are 160 farms in South Africa alone with 8000 captive lions [85]. All these lions need to be fed, which means many non-target animals are also killed for the lion hunting business. Farmed lions means there are always large numbers of lion cubs which start out in petting ranches, become tame, and are then sent to the kill zones [83, 85-86]. Tourists don't recognize that participating in lion or tiger cub petting and picture taking directly enhances the canned killing industry. The documentary ***Blood Lions*** shows the dark side of South Africa's captive lion breeding and killing industries [87].

Americans make up the largest fraction of the canned lion hunting customers, as much as 85% [74, 85-89]. Heads and hides are sent home to America while bones are sent to Asia. There is some hope of reducing the lion killing industry lately, however. In December 2015, the U.S. Fish and Wildlife Service listed the African lion as threatened under the Endangered Species Act, which prohibits the importation of lion trophies into the United States [89-90]. This would cut back on the customer base and render some of the lion farms in Africa unprofitable. It remains to be seen whether loopholes can be exploited or pressure from the hunting industry, the NRA, Safari Club, etc. will influence Congress to reverse this small bit of hope and protection.

WILDLIFE REFUGES

We know now, as we have always known instinctively, that animals can suffer as much as human beings. Their emotions and their sensitivity are often stronger than those of a human being. Various philosophers and religious leaders tried to convince their disciples and followers that animals are nothing more than machines without a soul, without feelings. However, anyone who has ever lived with an animal - be it a dog, a bird, or even a mouse - knows that this theory is a brazen lie, invented to justify cruelty.

- Isaac Bashevis Singer

When is a sanctuary, with all the implied asylum and protection that the word implies, not a sanctuary? When the U.S. Govt. labels it a Wildlife Refuge and opens it to hunting and trapping. A refuge, an inviolable place of safety and tranquility, is not a sanctuary anymore when the government and the hunting lobby get ahold of it.

The concept of "sanctuary" has a long and distinguished history. It was established at least by the 6th century in England where those accused of a crime or on the wrong side of a power struggle could find haven in a church. Political sanctuary is well known today when people are given asylum by one nation from persecution and arrest by another. Famous cases of political asylum are familiar. Tsar Nicholas II was denied sanctuary after the Russian revolution and executed in Siberia along with his entire family. In World War II, over 900 Jews sailing in the ship MS St. Louis were denied sanctuary first in Cuba then in the United States and wound up victims of Nazi death camps. More recently, Julian Assange has so far received sanctuary from extradition to Sweden from the Ecuadorian embassy in London, and Edward Snowden who leaked NSA surveillance documents is still living in asylum in Russia.

And who could fail to cheer when Quasimodo rescues the beautiful Esmeralda from torture and death by carrying her off to sanctuary in *The Hunchback of Notre Dame*!

The U.S. has had a number of "sanctuary cities" where undocumented people who haven't committed a crime are not reported or pursued [91]. These are localities that help protect undocumented residents from arrest and deportation and refuse to cooperate with immigration authorities. Legal wrangling takes the form of whether to deport such so-called illegals or give them a path to stay with possible citizenship. In 2015-16 the issue of immigration and human refuges came into stark exposure by the crisis of refugees from the wars in the Middle East. In 2017, the Trump administration ramped up the arrest and deportation of those designated illegal aliens, whether they had committed a crime or not.

The wildlife inside refuges can't ask for sanctuary and have no Quasimodo to carry them off to safety. The first wildlife refuge was created in 1903 by President Teddy Roosevelt, an avid hunter, when he became concerned with what appeared to be prolonged destruction of large numbers of wildlife. Since then, the list of designated wildlife refuges has grown to over 560 with hunting, fishing, and trapping allowed on a majority. Such activities were not allowed until the 1950's when consumptive use began on many of the refuges [92]. Hunting was expanded to more refuges in subsequent years, while trapping enthusiasts and the fur industry lobbyists managed to include trapping as another "recreational activity" in spite of the fact that endangered species like bald eagles and other accidental species including domestic dogs and cats are sometimes caught and killed as well as many other non-target species.

The actions of the NWRS (National Wildlife Refuge System) are all the more frustrating because its original intent, and the intent stated to this day, is to preserve and protect wildlife. Its mission statement reads "*To administer a national network of lands and waters for the conservation, management, and where appropriate, restoration of the fish, wildlife, and plant resources and their habitats within the United States for the benefit of present and future generations of Americans*" [93]. Nowhere in this mission statement does it state that the refuges should be managed for the benefit of hunters and trappers. The NWRS has had considerable success in restoring individual threatened species (even resulting in delisting of some) including plants, trees, insects, birds, antelope, rare deer, fish, marine life, and invertebrates [94]. Through its recent Cooperative Recovery Initiative (CRI) it has established well over 30 programs in all regions of the country to enhance species protection and recovery. Yet there is no way of getting around that hunting, fishing, and trapping are supported and encouraged on most refuges. The 1997 *National Wildlife Refuge System Improvement Act* designated hunting and fishing as **priority uses** in the refuges and "stipulated they receive enhanced consideration by refuge managers" [92]. Presumably hunters aren't supposed to shoot the same threatened species the NWRS is mandated to protect but since hunters occasionally mistake horses, cows, and people for deer, accidents are bound to occur.

Thirty million people visit wildlife refuges each year to watch wildlife and enjoy nature. Somewhere around 1.4 million go there to hunt and trap, yet by law wildlife managers are supposed to consider hunting a *priority*. A large majority of the public believes that trapping should be prohibited and a majority (88%) believe that priority should be given to wildlife and habitat protection [92], but the hunting and shooting lobbies have powers far beyond their numbers and Congress listens to donors and lobbyists rather than the public. As long as such a situation goes on, perhaps wildlife should be warned to stay away from federal Wildlife Refuges for its own safety.

National Forests and National Wilderness areas are also open to hunting, though National Parks are supposedly hunting-free. In a master sleight of hand, some National Parks have been re-designated as National Preserves or National Recreation areas and open to hunting. There has long been controversy regarding hunting near (not within) Yellowstone National Park, the granddaddy of them all, when elk, bison, wolves, bears, or moose may wander outside the park boundaries into the sights of eager hunters waiting to shoot them.

And matters seem to be getting worse. The Republican Congress and President Trump passed a resolution in mid-2017 permitting even more killing on wildlife sanctuaries, this time on 16 National Wildlife Refuges in Alaska, allowing hibernating bears and wolves with their pups to be shot in their dens, the chasing and shooting of grizzly bears from helicopters, and the use of more leghold traps and snare traps to kill bears and other predators, all to increase elk, caribou, and moose populations for hunters. Alaska's wildlife sanctuaries are now officially game preserves, for the benefit of the NRA and the Safari Club, and the delight of many hunters.

EXTINCTION

Man may for a time deplete or erode the soil, denude the forests, pollute the streams, kill the wildlife, make numerous species extinct, spread atomic radiation, and even poison his own food with insecticides and herbicides. But what he sows he will at last reap; his karma will accumulate destructively. To live, and live creatively, man must deepen his sense of the holiness of nature.

- Howard L. Parsons

Poaching is one of the main causes of animal endangerment and extinction. This act of killing large numbers of wildlife, where a considerable fraction of the remaining population of a species may be destroyed each year until all are gone and the poachers move on to another species, is exceedingly destructive. For short term profit, poachers are willing to drive a species like the elephant or tiger to extinction, species that have evolved on our shared planet over countless millennia.

However, there are other forces driving species to extinction as well. Habitat destruction is probably the main driver, as the human population grows without restraint and the demand for farmland, building land, development, and recreation continues to expand. Overall, wildlife is often reduced to a mere 5-10% of its former range, and even places set aside to preserve wildlife like the refuges, as we have seen, become hunting and trapping preserves instead. A second factor that threatens predator populations in particular both on land and at sea is decimation of prey species. In a well-balanced ecosystem, predator and prey species have a symbiotic relationship. It's nature's way; each keeps the other healthy and strong, with a resilient gene pool. Damage to either one damages the other. Humans are doing a good job of creating that damage.

Total destruction of a population isn't even necessary to accomplish irreparable damage. Once a species is reduced below a certain number it becomes *biologically* extinct; the gene pool has become too restricted to maintain a robust, healthy population able to ward off diseases and maintain long term health and strength. Over time, it will be unable to survive in the natural world it must face every day, and it may become truly extinct. Members of a species may be reduced so greatly in number that they are unable to find a suitable mating partner, causing eventual die-off. For predator species, reduction in the population of a critical and necessary prey species is another way of making their survival difficult.

Hunting, the legal form as well as the illegal – poaching - is another major cause of species extinction, and if you add the two together, may very well rival habitat destruction as the number one cause. The roster of animals gone forever due to hunting is long and growing. David Day has written a remarkable compendium of extinctions titled *The Doomsday Book of Animals* [95] that only covers the time period between 1680 and 1980, so doesn't include the mastodons, mammoths, sabre-tooth tigers, and other iconic species that come to mind when people think of extinction. If he had written the book in the 21st century, the list would likely have been more than doubled, just in the past 30-40 year period. The list of species made extinct by hunting before 1980 includes:

Elephant Bird, 10 feet tall, 1100 pounds, evolved over 60 million years, gone by 1700.

Giant Moa, 13 feet tall, 600 pounds, evolved over 100 million years, gone by 1850.

Dodo bird, found on the exotic islands of the South Pacific, gone by 1680.

Passenger Pigeon, once numbering in the billions and accounting for 40% of all birds, exterminated in the late 1800's. The last one died in a zoo in 1914. In a single hunting party in Michigan in 1878, hunters destroyed an estimated 1000 million (1 billion) birds.

Great Auk extinct by 1844; Spectacled Cormorant, extinct by 1850. Eskimo Curlew, a major insect predator and a godsend to farmers, hunted to extinction for fun by 1900.

Raptors, the fierce and spectacular birds of prey, are among the most critically endangered creatures in the world [95]. Today, they consist of the eagles, falcons, hawks, vultures, condors, and owls. In the past, some raptors and many other types of birds became extinct through habitat destruction or by the introduction of invaders such as cats, rats, mongooses, dogs, rabbits, weasels, or pigs, and sport hunting hastened or finalized their demise.

The list of bird species exterminated primarily by hunting continues:

Quelili, a large brown hawk, extinct by 1900. Mauritian Barn

Owl, extinct by 1700. Carolina Parakeet, America's only native parrot, extinct by 1914. Broad-billed Parrot, the largest of the parrot species, extinct by 1650. Norfolk Island Kaka, another parrot-like species, extinct by 1851. White Gallinule, a rail bird species, extinct by 1847.

Heath Hen, extinct on the mainland by 1830, extinct on Martha's Vineyard by 1932. New Zealand Quail, extinct by 1868. Pink-Headed Duck, extinct by 1924. Labrador Duck, extinct by 1871. Crested Shelduck, extinct by 1916. American Ivory-Billed Woodpecker, extinct by 1972.

Hundreds of bird species became extinct through a combination of introduced invaders, specimen collecting, habitat loss, and hunting, including rails, exotic Hawaiian island species such as honeyeaters and honeycreepers, and species of thrushes, warblers, finches, grosbeaks, wrens, starlings, quails, grouse, woodpeckers, and ducks.

For mammals, all the forces driving extinction described earlier as well as poaching still remain major factors in threatened and endangered species and extinction. Endangerment and extinction of many species are major threats today, not just in years gone by. Many iconic species in the early 21st century, including tigers, rhinos, some bears and wolves, some species of whales, and many others, may very well be gone in a few decades to come.

The wolf in its several subspecies has been the most persecuted of the natural predators (a distinction that has now migrated to coyotes). Attempts have been made to exterminate wolves all over the world. There are no wolves left in many parts of Europe and they are also gone from most of their former range in North America. Since hunting wasn't fast enough, the widespread use of the fast acting poison strychnine began in the 1800's. Not only wolves died very painful deaths but so did coyotes, foxes, weasels, cougars, bears, skunks, badgers, ferrets, squirrels, raccoons, eagles, bobcats, ravens, occasionally horses, buffalo, and antelope. Indians were also occasional victims, as were cattle, dogs, and children. In 1905, a veterinarian deliberately introduced sarcoptic mange into the wolf population, which quickly spread to cattle and dogs and can also infect people. The act should be considered a form of torture, causing intense and uncontrollable itching. Mange is highly contagious.

Hunting of the few remaining wolves and other predators remains to this day. The list of extinct predators and other mammals includes:

Newfoundland White Wolf, pure white, extinct by 1911.

Kenai Wolf, perhaps the largest of all wolf species, extinct by 1915.

Texas Grey Wolf, extinct by 1915.

New Mexican Wolf, extinct by 1920.

Great Plains Lobo Wolf, once among the most numerous, extinct by 1926.

Southern Rocky Mountain Wolf, inhabited Nevada Utah, and Colorado, extinct by 1940.

Cascade Mountains Brown Wolf, extinct by 1950.

Japanese Wolf, the world's smallest wolf, extinct by 1905.

Florida Black Wolf, native of Alabama, Tennessee, Georgia, and Florida, extinct by 1917.

Texas Red Wolf, extinct by 1970.

Antarctic Wolf (or Warrah), studied by Darwin, extinct by 1876.

Atlas Brown Bear, Africa's only bear, extinct by 1870.

Mexican Silver Grizzly, poisoned, trapped, shot, extinct by 1964.

Since 1850, 22 subspecies of grizzly bears throughout the U.S. were hunted to extinction.

Kamchatkan Black Bear, 1500 pounds, extinct by 1920.

Barbary Lion, Tunisian Lion, Algerian Lion, all extinct by 1922.

Cape Lion (South Africa), extinct by 1865.

Bali Tiger (Indonesia), extinct by 1937.

Arizona Jaguar, extinct by 1905.

Auroch (giant bull famous in mythology), extinct by 1627.

Eastern Bison, extinct by 1825; Oregon Bison, extinct by 1850; Caucasian Wisnet (large buffalo), extinct by 1925.

Blue Buck; Bubal Hartebeast; Rufous Gazelle; Schomburgh's Deer; Dawson's Caribou; Eastern Elk; Badlands Bighorn Sheep; Portuguese Ibex, all extinct by 1932.

Wild Horses: Quagga, extinct by 1883; Syrian Onager, extinct by 1930; Tarpan, extinct by 1887. Pictures are available in [84].

Steller's Sea Cow, discovered in 1741, extinct by 1767.

Caribbean Monk Seal, extinct by 1952.

Sea Mink (New England), extinct by 1880.

Several of the great whales, many species reduced to 10% of their former population, may be close to biological extinction.

Tasmanian Wolf, also known as Thylacine, a unique species with the head of a wolf, stripes like a Zebra, tail like a possum, and pouch like a Kangaroo, extinct by 1933. Pictures of this remarkable species are available in [96].

Toolache Wallaby, most beautiful of the Kangaroo family, extinct by 1940.

Rodriguez Saddleback Tortoise, extinct by 1800.

Mauritian Giant Tortoise, Domed Giant Tortoise, High- fronted Giant Tortoise, all extinct by 1700.

Reunion Giant Tortoise, extinct by 1773.

Galapagos Giant Tortoises, 4 subspecies extinct by 1957.

This list, which is only a fraction of the species that have become extinct mainly or exclusively due to hunting, can be supplemented in the hundreds by species that are threatened or endangered. An example might be the tiger, where perhaps 3000-3200 remain in the wild and the Asian market will likely doom these as well. Once the population is reduced in number to a critical point, they could become biologically extinct, unable to reproduce in the wild, maintain a varied gene pool, and survive as "true tigers." The many captive tigers kept in sanctuaries and the appalling tiger "farms" may keep the species alive for some years, but they would not survive as true "wild tigers" as nature made them, and even if tiger re-introduction programs were established, who would be available to train them in the ways of the wild?

The Asian market is also endangering the pangolin, a unique armadillo-like creature killed for its scales and meat, and has almost driven the world's smallest porpoise, the vaquita, to extinction as a by-product of fishing for the endangered totoaba [97]. The Asian market for food, so-called traditional medicines, and status symbols is killing elephants, tigers, bears, pangolins, vaquitas, totoabas, rhinos, sharks, leopards, sea turtles, and several species of rare birds. In the U.S., next to China the biggest consumer of endangered wildlife, there is a thriving illegal trade, sometimes for exotic pets but more often as exotic meat, including gorillas, chimpanzees, monkeys, snakes, baboons, antelope, turtle, and tortoise [98].

The list of species alive today that could be gone in a few generations includes countless invertebrates, reptiles, birds, fish, amphibians, sea turtles, the great cats, bears such as grizzly and polar bears, gorillas and other primates, whales and other sea creatures, and many types of small mammals. As the human population grows and habitat shrinks, many of these species have no place to go. Global warming will greatly exacerbate the problem as will ocean and land pollution, but hunting, poaching, and habitat destruction will remain as major causes of extinction and ones that could have been most easily stopped, if humankind only cared enough.

CHAPTER 2
Helping People

HUNTING ACCIDENTS, INJURIES, FATALITIES

The time will come when men such as I will look upon the murder of animals as they now look upon the murder of men.

- Leonardo Da Vinci

Our task must be to … widen our circle of compassion to embrace all living creatures and the whole of nature in its beauty.

- Albert Einstein

It's obvious that hunting involves the use of deadly weapons. One doesn't kill a bear with a slingshot or a deer with a pocket knife. Every year, and in the fall in particular, men, women, and children traipse through the forest carrying rifles, handguns, shotguns, or bows and arrows for the purpose of killing animals. Accidents due to loaded firearms are almost predictable; a hunter trips over a log or bush and discharges his weapon into the back of the guy ahead or into his own body, or hunters leaving their vehicle and reaching for their weapon, pull the trigger of a loaded rifle by mistake. There are even incidences of dogs shooting their owners as the enthusiastic canine, out for an exciting day in the woods, catches their leash in the trigger region of their master's weapon. Strangely enough, there are more accidents, and perhaps more fatalities, from bow hunting, as bow hunters fall from carelessly mounted tree stands or fail to take precautions as they climb the tree to take their place on the stand [1].

Every year, there are over 100 fatalities in hunting accidents and at least 10 times as many non-fatal accidents. The National Shooting Sports Foundation in their 2013 and 2014 reports list 6,759 hunting accidents with firearms in 2013 and 7,302 in 2014 [2], but adding tree stand accidents more than doubles that number. Even exposure to the elements constitutes risk, dangers which aren't included in National statistics, such as the woman who lost both legs to frostbite while elk hunting in Oregon, or the duck hunters who drown. A look at the headlines from newspapers and journals around the country gives some idea of the horrors of these accidents. These are easily found on the internet, but for convenience, C.A.S.H. (Committee to Abolish Sport Hunting) has compiled many of them by year and location [3]. A small sampling of hundreds of headline entries includes:

MI: Man fatally shoots self while hunting coyotes.
TX: 'Highly respected' Connecticut trooper dies in Texas hunting accident.
AK: Three moose hunters killed in boating accident.
KS: Hunter drowns at Kansas Lake.
AR: Editor dies in fall from tree stand.
AL: Teen shot by brother in hunting accident.
WA: Body of missing hunter found in eastern Washington.
MS: Father kills 16-year-old son in hunting accident.

IA: Worthington man injured by self-inflicted gunshot wound while trapping.

KS: Winchester man killed in Jefferson County hunting accident.

OH: 13-year-old injured in hunting accident.

MI: Gun hunter shoots bow hunter.

MN: Deer hunter dies in MN after 20-foot fall from tree stand.

IA: Teen impaled through neck in Fayette County hunting accident.

CT: 14-year-old shot by hunter at Salmon River State Forest.

WI: Hunter dies after tree stand fall in Crawford County.

NE: Man deer hunting near Weeping Water dies in rollover crash.

GA: Hunter critical after 30 foot fall from deer stand.

NY: Hunter shoots man while attempting to unload gun in Calverton.

WY: Young father dies of altitude sickness while hunting in Big Horn Mountains.

DE: Hunter shot with bow and arrow in hunting accident.

IN: Indiana hunter dies trying to remove tree stand.

PA: Hunter accidentally kills self with crossbow.

VT: Girl shot in stomach by hunter-father loses kidney.

IN: Indiana waterfowler shot by hunting dog named "Trigger."

OR: Father accidentally shoots, kills son while hunting in Eastern Oregon.

UT: Boy, 10, accidentally shot, killed during Utah hunting trip.

NY: Hunter fires shotgun at squirrel, hits fellow hunter.

MT: Bow hunter missing near Wisdom found dead.

WI: Slinger man dies after falling from tree stand in Washington County.

MI: Porcupine hunter attacked by black bear in Michigan woods.

> TX: Man shoots 9 year-old grandson in fatal hunting accident.
>
> ME: Man accidentally shot by wife while turkey hunting.
>
> MO: Man mistakes, shoots fellow hunter for turkey.
>
> PA: Hunter who allegedly paralyzed woman heads to county court.
>
> VA: 13-year-old shot in Scott County hunting accident.
>
> MS: October squirrel hunting accident blinds Harahan hunter.
>
> IA: 12 year-old Traer girl killed in weekend hunting accident.
>
> PA: Newborn baby shot in head in Pennsylvania 'hunting accident.'

Yes, you're reading that last one right; a newborn baby cradled in his parent's lap inside their house was shot and blinded by a stray bullet fired by a deer hunter.

Wives have shot husbands, husbands their wives, fathers their sons, sons their fathers, fathers their daughters, daughters their fathers, brothers their brothers, brothers their sisters, grandfathers their grandchildren. Hunters mistake each other for deer, bears, and turkeys, shoot cows, horses, and cars thinking they are deer, drown in boating accidents while hunting waterfowl, die from exposure in freezing weather, become lost in the forest and are sometimes not found, even become injured or die in avalanches.

It seems obvious that if hunters weren't trying to kill wildlife, these injuries and fatalities wouldn't have happened, and the deaths and injuries of thousands of people would be prevented. Hunting proponents like to claim that 60,000 people die in car accidents every year so hunting is really much safer than driving. They gloss over the point that a few hundred million people drive cars at high speeds for many hours a week, while a few million hunt in the woods for a few days a year, so looking at it in terms of person-hours, car driving is much safer than hunting [4], just as commercial airplane flying is safer than driving a car in terms of fatalities per passenger mile.

LYME DISEASE

When a man wantonly destroys the works of man, we call him a vandal. When he wantonly destroys one of the works of God, we call him a sportsman.

\- Joseph Wood Krutch.

When a man wants to murder a tiger, it's called sport; when the tiger wants to murder him, it's called ferocity.

\- George Bernard Shaw

One of the worst infectious diseases that has emerged in centuries is Lyme disease. It's not an instant killer like ebola, which can end a person's life in a few days. Instead, Lyme disease, if misdiagnosed or left untreated, can cause a person to suffer debilitating symptoms for decades. This illness which is becoming an epidemic as the years go by is caused by a bacteria shaped as a spirochete, transmitted by far mostly by ticks, though there is some thought that mosquitoes may occasionally also transmit the bacterium. If one is lucky enough to notice the Lyme tick bite from its characteristic rash or other less common symptoms and take a regimen of antibiotics soon after the infection occurs, it can be cured rapidly. However, in many cases there are no tell-tale symptoms and months may go by before the illness appears. At that stage, the disease can become permanent, known then as chronic Lyme disease (which the AMA and insurance companies "officially" deny exists) and it can cause crippling arthritis, brain fog, paresthesia, constant headaches, palsy, chronic fatigue, muscle ache, inability to work, depression, memory loss, anxiety, agitation, meningitis, kidney damage, cardiovascular problems, and a number of other ailments, including in rare cases seizures and strokes [5-6]. Chronic Lyme Disease may be very difficult or nearly incurable at that stage; though the medical community recommends long (months) treatment with intravenous antibiotics, but even these don't always work. The CDC now believes that there are 329,000 *new* cases of Lyme disease per year [7], and that number is increasing.

Whole lives have been ruined by this disease. To make matters worse, the same tick that transmits the Lyme bacterium also often carries one or more co-infections such as anaplasma, mycoplasma, bartonella, babesia, Q-fever, or ehrlichia. Tests to detect which of these a patient has contracted are difficult and inaccurate, and cures are difficult and often elusive. Of all these co-infections, Lyme disease is far and away the worst because of its tenacity once it's taken hold, and it is infinitely better not to contract it in the first place.

At the beginning of the life cycle, the tick is in the larval or nymph stages and the preferred host is the field mouse, which appears to tolerate the infection but harbors the bacterium and allows it to multiply. The mouse acts as a conduit; it becomes infected by infected ticks, and then passes on the infection to other ticks. Once the tick enters the adult stage the preferred host is the white tailed deer, though ticks continue to infest small rodents: chipmunks and squirrels as well as the field mouse. Deer are critical for the survival of the ticks [6]. It has been shown that the tick population tracks the deer population, and when there are less deer, there are less ticks and consequently less Lyme disease [6-7]. In one CT study, when the deer population was reduced from 80 to 13 per sq. mile, the number of new Lyme cases dropped from 20-30 per year to 2-5 per year [7]. Other studies found less correlation, possibly because ticks had switched to other hosts, including humans. Large ungulates like moose and elk have been found to be infested with hundreds if not thousands of ticks on a single individual.

Wildlife management practices that encourage greater deer populations therefore bear large responsibility for greater numbers of ticks and greater risks of humans contracting these diseases. In the past, most of the ticks carrying Lyme and the co-infections were found in the forests, the exact locations where hunters spend their time during hunting season. However, high tick populations are now found in lawns, gardens, back yards, and any property where chipmunks, squirrels, and mice are found and where deer have passed through, making the general rural and suburban populations at great risk as well as hunters. State wildlife agencies set out throughout the 20th century to increase deer populations to satisfy hunters, by instituting "buck laws" where mostly or only male deer could be killed (more on this later), and by controlled burns of forests to create favorable browse habitat for deer. (Such burns also cause the destruction of countless small animals and birds [8].) Policies were also put in place to exterminate natural predators which could have controlled the deer population as nature intended but which competed with hunters; these anti-predator policies continue today. These agencies, arms of our state governments, are therefore responsible directly or indirectly to some degree for the spread of Lyme disease and the growing number of people who suffer consequent debilitating diseases.

To put it more bluntly, killing animals, male deer and the predators that might have controlled their population naturally, is indirectly responsible for a great deal of human danger and suffering, and the risk to people is increasing every year. Arguments have been made that bringing predators back to impact the deer population, and removing males and females in equal numbers, would be beneficial to reducing the spread of Lyme disease and the co-infections. "It's {Lyme Disease} now the single greatest vector-borne disease in the United States, and it's expanding on a really epic scale… Predators can really regulate infectious disease and actually protect us," D. Buttke, National Park Service [9].

DEER-CAR COLLISIONS

I will not kill nor hurt any living creature needlessly, nor destroy any beautiful thing, but will strive to save and comfort all gentle life, and guard and perfect all natural beauty upon the earth.

- John Ruskin

It is our duty as men and women of God's redeemed creation to try not to increase the suffering of the world, but to lessen it. To get rid of blood sports will be a great step towards this end.

- Leslie G. Pine

At the beginning of the 20th century, the white-tail, black-tail, and mule deer were almost endangered species. They had been nearly extirpated by hunting. There were no significant state wild-life agencies at the time and no hunting regulations. Deer of both genders were hunted equally, and the deer population of 35 million in the year 1700 had been reduced to less than 500,000 in 1895 [10-11]. In 1900 there were less than 1 deer for every 7.6 square miles, and it would have been rare to see one. By the year 2000, there were over 38 million [10], a 100-fold increase, and it became common to see deer eating the local flowers, shrubs, and plantings. What also took place as the deer population increased and the number of cars on the road also increased was ever increasing instances of deer-car collisions, responsible for 175-200 deaths and 10,000 injuries per year [12]. From insurance reports, there are 1.2-1.5 million deer-car collisions per year costing $4 billion, or $3,300 in car repairs on average for each car [13]. In 1980, there were 19,000 deer-car collisions in Michigan alone. In 1978 there were 28,700 such collisions in Pennsylvania [Chap.1, ref 2].

In the early 1900's, when the deer population was so low, wildlife agencies were born to increase the deer population for the benefit of hunters. The numbers above demonstrate the enormous success of these attempts. What methods were used to effect this dramatic change? Eliminating natural predators was one, manipulating the habitat to encourage more deer food, "browse," was another. But the biggest method to increase the population was hunting.

What? Increased hunting that might logically lower the population actually increased it? How can that happen? Countless pundits have fooled the public into thinking that more hunting can reduce deer numbers [14], save their plants, and prevent those thousands of collisions, injuries, and deaths. The answer is that it's not whether you hunt, but *how* you hunt. Wildlife agencies started restricting hunting to male deer, the so-called "buck laws" [11]. The artificial ratio of male to female deer that they caused has the effect of increasing the deer population, a result of the dynamics of "mother nature."

Normally the population of any wildlife species is determined, on the average, by the carrying capacity of the habitat the wildlife occupies. The crucial factors are the food supply, the water supply, the existence of predators, and the availability of some type of shelter from extreme elements. The death rate from all causes matches the birth rate if all is in balance; if more members of the group die, there are more resources for the remainder and the number of newborns increases until the carrying capacity is reached again. If all these dynamic factors are constant, the population, on average, will also be constant.

To use an example, suppose there are 100 deer, the carrying capacity of a particular local environment. With a normal ratio of 1:1 male to female, there will be 50 *does* and 50 bucks. With a longevity of 5 years, from hunting and natural causes, 10 bucks and 10 *does* would die during the year, and the 40 remaining *does* would give birth to around 20 fawns as determined by nature and the carrying capacity. The population would remain stable at 100. Now suppose instead that more bucks are killed so the population is reduced to 40 *does* and 20 bucks, perhaps with the excess bucks killed by hunters. The over-abundant food supply causes the *does* to give birth to multiple fawns, as many as two, but for discussion suppose the average is 1.5 fawns per *doe*. In the springtime, then, there will be 40 *does*, 20 remaining bucks, and 60 fawns, for a total of 120 deer. Hunting mainly bucks has increased the deer population due to nature's increased birth rate. This increased birth rate is nature's response to population dynamics and it applies to many species besides deer. The excess multiplies in subsequent years, and the deer population continues to increase. Add habitat manipulation, which *increases the carrying capacity*, and the deer population increases even faster.

The tragedy is that as the deer population rises, so too does the number of deer-car collisions and consequent fatalities and injuries. People are being hurt as a result of animals being killed. This effect of hunting is contrary to what the public is led to believe, but it's standard practice for state wildlife managers. Do these agencies realize that this suffering and deaths are a result of their management practices? Judge from their own words.

> "We will attempt to increase the number of deer until we experience high incidences of deer-car collisions, depredation of agricultural crops becomes intolerable, and/or the effects on deer habitat begin to result in deterioration." Terry Moore, regional wildlife manager for the New York State Department of Wildlife, quoted in [8] and [11].

> "The most visible weakness in the assertion that hunting is necessary to control deer populations is that it has largely failed to do so over the last two decades. ... Just because

deer are being killed doesn't mean that deer populations are being controlled." [15].

"The long-range objective of our deer management program in NYS has been to provide the largest possible harvest of antlered deer compatible with land use and deer herd health ...It ensures that deer will be in good physical condition and can grow and reproduce at optimum rates." [16].

"New York State has rapidly seen its whitetail deer herd grow from modest thousands to more than an estimated million animals today. Every county in the state has a good population of deer, and many have too many. Management techniques by DEC have changed from how to increase our herd to how to keep it under control and successfully manage it for future sportsmen." [17].

"In 1912, New York State passed the "buck law" which restricted hunters to killing only bucks. The remaining *does* were then free to reproduce at their maximum capacity due to less competition for available resources, such as food. According to The Conservationist, the primary objective of this law was to foster the greatest population growth by protecting all fawns and adult females from hunting. This regulation contributed greatly to the increase and spread of the deer population that took place in the state during the next thirty years." [11].

"Ideally, if the desired number of antlered and antlerless deer are taken each year, the population will comprise the highest number of breeding females and lowest number of adult males that collectively can be supported on the critical winter range. As a result, a maximum fawn crop will be produced each summer." [11, 16].

"Don't lose sight of the purpose of the (DRIP) program. When the DNR decided several years ago to try and increase the herd to about one million animals, we knew the auto collision rate and crop damage would rise." [11, 18]. (DRIP = Deer Range Improvement Program. DNR = Department of Natural Resources.)

"Add to this the disproportionate number of licenses issued to kill male deer and you have a major increase in numbers. Killing male deer increases the herd by causing the females' internal reproductive mechanisms to go haywire, resulting in the births of twins and triplets. ... That makes the hunting community solely responsible for the increase in deer-car collisions" [19].

"Deer-car collisions: an inevitable by-product of recreational hunting. The states with the most hunters have the highest numbers of deer and deer-car collisions. The number of deer-car collisions has risen steadily along with the size of the deer herd in New York State" [11].

"The Wildlife Division of the Department of Natural Resources (DNR) in Michigan increased the deer herd from 400,000 to one million by clear cutting 1.3 million acres of state forest to create deer browse. According to officials, this was done "because a forest managed by nature cannot produce a fraction of the deer needed by half a million hunters. A side effect of this deer production program was that 171,207 of these deer starved to death during the 1978 - 1979 winter" [11].

According to the National Highway Traffic and Safety Administration, most car/deer collisions happen during hunting season. It is not difficult to understand why the hunting going on in the woods would send deer out onto the roads in a panic.

What can be done to reduce the deer herd to acceptable levels according to what nature intended and save people's lives and injuries from deer-car collisions? Leave the herd alone so it can recover according to the carrying capacity of the local environment. Leave the predators alone to restore nature's balance. There will still be some collisions, since the deer population won't be zero, but deer won't be running out of the forest onto roads after being terrorized during hunting season.

Cut back on hunting, save animals and save people too; what a radical thought!

EXTINCTION

What is the nature of a species that knowingly and without good reason exterminates another? How long will man persist in the belief that he is master of this earth rather than one of its guests? When will he learn that he is but one form of life among thousands, each one of which is in some way related to and dependent on all the others? How long will he survive if he does not? Whatever the nature of the Creator, He surely did not intend that the forms on which He bestowed the gift of life should be exterminated by man.

- George Small.

Many people may wonder why we should care if various species become extinct. After all, the dinosaurs went extinct, the saber-tooth tiger disappeared, all those great birds, bears, and wolves listed earlier are gone. Even many hominin species have gone extinct for various reasons over the millennia. We humans don't have to worry about these things, right? We don't have to worry about becoming extinct ourselves, right?, from warfare, emerging diseases, disappearing resources, environmental deterioration, political upheaval, and other ongoing issues, right?

The earliest known hominin species on the human ancestral tree is homo habilis, a 4 foot tall, 70 pound hominid that lived in Africa 2 million years ago, and became extinct around 1.4 million years ago. Then there was homo rudolfensis, also from Africa, who became extinct 1.8 million years ago. This was followed by homo erectus from Africa and parts of Asia who became extinct 140,000 years ago, homo heidelbergensis from Europe, China, and Africa who disappeared 200,000 years ago, homo floresiensis from Asia, extinct about 50,000 years ago, and homo neanderthalensis from Europe and Asia that exited the earth around 40,000 years ago.

Now *we* are the latest on the evolutionary tree: homo sapiens, evolved over the last few hundred thousand years. We don't have to worry that we'll become extinct like all our ancestors, in spite of nuclear weapons, ubiquitous and ever increasing chemicals and pollution, rain forest and temperate forest destruction, decreasing water resources, religious warfare, intolerance, and hatred, terrorism, racism, political warfare with its land mines, cluster bombs, and poison gas, biological warfare, chemical warfare, unbounded population increase, acid rain, climate change and all its consequences, saber rattling between East, West, and Middle East, cyber warfare where nations practice at destroying each other's infrastructure, and extinction of other species at the rate of 50 a day. Technology, common sense, and homo sapiens' intelligence will come to the rescue. Hooray for us!

Like all other species on Earth, nature establishes a carrying capacity for homo sapiens, based on the amount of resources (food, water, shelter, waste removal) necessary to sustain our species in status quo, replenishing resources and eliminating waste at a rate which ensures good health. Unfortunately, we seem to be unaware of this concept or choose to ignore it, but population growth, loss of topsoil and forests, widespread pollution, damage to the oceans, and above all our wholesale destruction of countless vital species, are all taking us in the wrong direction, reducing the carrying capacity while increasing demand. This is a blueprint for eventual disaster at some point in the future, a disaster visited particularly upon our grandchildren and their descendants. As a species, we seem to be incapable or unwilling to diverge from this path.

Needless to say, respecting nature and preventing extinction, of plants, animals, and invertebrates, on which we all collectively depend, would have enormous benefits for people. Helping animals is helping people.

The effect on people of species extinction is an enormously complex question that involves ecology, biological systems, and biodiversity. A few simple examples have already been mentioned earlier of synergy between species and its effects: the sea otters-urchins-kelp beds; the Yellowstone wolf-elk interaction, and the starfish-shrimp-Great Barrier Reef connection. But all these affect people only *indirectly*. Other effects of species loss affect people much more directly.

The iconic example of extinctions that affect humans is the disease-medicine connection involving plants and sometimes other life-forms. There are many examples of medicines that originated from nature that continue to save human lives. One of the most ubiquitous and destructive illnesses throughout history is tuberculosis, shown to be present from artifacts dating 20,000 years ago. It was also known as "consumption," and killed countless human beings over the years. Such personages as Robert Louis Stevenson, Emily Bronte, Edgar Allen Poe, and Frederick Chopin suffered from it.

The first effective treatment for the disease was the antibiotic strep-tomycin, developed from soil bacteria, along with 15 other effective antibiotics, discovered by Selman Waksman and his students. Fortunately for humanity, pesticides and herbicides hadn't yet been sprayed everywhere, driving the soil organisms to extinction, before this enormously important discovery from one of nature's tiny invertebrate species was made.

The list of key medicines that have saved countless lives but *could have been lost* from deforestation, herbicides, agriculture, or thoughtless human activities includes the cancer treatments vincristine and vinblastine from the periwinkle flower, quinine to treat malaria from the cinchona tree or more recently artemisinin from sweet wormwood, digitalis to combat heart failure from the foxglove plant, taxol from the Pacific Yew tree used to fight breast, lung, and ovarian cancer, and hundreds of others. Long lists of antibiotics used to treat previously fatal infections (plague, typhus, scarlet fever, diphtheria, pneumonia) have been developed from species of fungi and bacteria. In the present day, the increasing antibiotic resistance of disease pathogens makes previous antibiotics less useful, and new antibiotics to replace these will come from plants, beneficial bacteria, molds, and fungi, *as long as we haven't driven these critical unknown future sources to extinction.*

Start with the lowest of the low: bacteria and insects. Should we care if bacteria become extinct, at least some of them? Bacteria cause many diseases, yet without bacteria we wouldn't have all those antibiotics mentioned earlier and many people would have lived shorter lives. Another key role of bacteria is in helping to digest food in the human (and animal) intestines, strengthening the immune system, and extending life [20]. *Beneficial* bacteria also control yeast and fungal infections and minimize *harmful* bacteria in the gut.

What about bugs, insects, and other invertebrates; we could do without those, right? According to scientific studies, 98% of all insects provide biological services that keep the environment healthy [21]. The effect of bees in pollinating plants is well known, but less well known is the effect of insects, plants, and bacteria in decomposing the waste that humans and other animals create. All the sewage, all the dead and decaying bodies, the leaves that fall in autumn and the trees and plants that die in the forest, are returned to the environment, recycled, by these unsung invertebrate heroes. "These roles provide immense benefits to humans and we depend on the pollination of crops by insects for our food and livelihoods. Invertebrates are also a source of food for many species including humans and the diets of some birds, fish and mammals are entirely dependent on them" [22].

It is estimated that human beings are dependent on 40,000 different species of invertebrates in our daily lives [23]. Of course, people don't think of that fact, let alone know what these thousands of species are, but they keep the planet and us humans healthy. Driving many of them to extinction, by pesticides, herbicides, habitat destruction, and the increasing effects of climate change, is detrimental to humans in ways we don't even know yet. These tiny, familiar species: worms, spiders, ants, insects of all kinds, mites, moths and butterflies, and countless others literally keep the planet healthy, and without these little critters, vertebrates such as human beings wouldn't survive very long. "Without invertebrates, the global ecosystem would collapse, and humans and other vertebrates would last only a few months, and the planet would belong mostly to algae and fungi. People hate the myriad small, spineless creatures on which their very existence depends. They maintain the soil structure and fertility on which plant growth and thus all higher organisms depend. They cycle nutrients by consuming decaying matter, pollinate crops, disperse seeds, keep populations of harmful organisms under control, and eliminate wastes" [24].

What can we do to thank these nameless organisms that literally keep us alive and protect and preserve our environment? Cutting back on the indiscriminate use of pesticides and other chemical pollutants would be a good start.

Phytoplankton, the almost microscopic plant life covering the surface of the sea, are the bottom of the food chain for all the species living in the ocean. They also capture CO_2 from the atmosphere and generate more oxygen than all the trees and plants on land combined. They help people by reducing greenhouse gases and providing oxygen to breathe. But the population of phytoplankton is in steep decline, most likely from global warming that reduces available nutrients by changing the ocean upwelling dynamics [25].

Salamanders and frogs, two freshwater invertebrates that include mosquito larvae in their diet, are in decline due to pesticide proliferation. Mosquitoes carry diseases that kill millions of people each year, and predator species like frogs provide a first line of defense on our behalf, yet we ignore this benefit and continue practices which diminish their populations. Some species of frogs, salamanders, and other amphibians are in serious danger of extinction [26-27], and humans will be the worst for it.

The list goes on and on of plants, animals, and invertebrates that are beneficial to human well-being, sometimes for their aesthetic majesty like whales, tigers, and elephants, sometimes more directly like frogs that eat mosquito larvae, tiny coral animals that create barrier reefs that protect against waves that otherwise damage coastlines, plankton that fight CO_2 and create oxygen, forests that protect topsoil and filter water, plants and bacteria that provide antibiotics or promising cures for cancer, predators that create healthy ecosystems, insects that pollinate our flowers and food, and countless bacteria, worms, and insects that turn our waste into nutrients and make continued life possible. We owe so much to these non-human life forms, but we damage them every chance we get in the name of profit, progress, or just human being's rights to treat nature any way we want [28], including driving many of these life forms to extinction.

It's not as if humanity hasn't been warned. For years biologists and naturalists have been warning about the consequences of species loss, damage to biodiversity, and extinction [29-30]. These people don't have a hidden agenda to damage capitalism and human progress; they've simply been warning that in the long term, human survival depends on a healthy planet and that in turn depends on healthy ecosystems. As species disappear, we don't know what specific damage will be caused in the future, but we do know that damage of some sort is inevitable. Sometimes there are keystone species that sit at the apex of a cascade of events, such as the wolves that affect the elk of Yellowstone, that in turn affect the aspen, that affect the wildlife populations and that affect the frequency of wildfires, that affect the burning of homes and tragic loss of firefighter lives. Or the sea otters that determine the urchin populations that damage the kelp that support the fish and other sea life that feed people. The best hope for the health of the planet doesn't lie in religion or expanding human populations or more chemicals to counteract the effects of other chemicals or fracking to replace one fossil fuel with another; it lies in working in harmony with nature, in biodiversity and large, healthy ecosystems that allow countless interacting species to live harmoniously without human interference [29-30]. Habitat loss, biodiversity loss, extinction, and hunting and poaching, are ultimately detrimental to the long term health of homo sapiens.

In a hundred years, history will hardly remember the fight against bin Laden or Al- Qaeda, but countless future generations for millennia will grieve for the extinct elephants, rhinos, tigers, whales, lions, gorillas and so many others that we could have saved. They'll even grieve for the tiny invertebrates whose effects on the health of the planet could have saved so many human lives. What those future generations will think of the present ones we can only imagine.

POACHING, PARK RANGERS, AND TERRORISM

*Until we establish a felt sense of kinship between our own
species and those fellow mortals- those "other nations"
as Henry Beston put it – who share with us the sun and
shadow of life on this agonized planet, there is no hope
for other species, there is no hope for our environment,
and there is no hope for ourselves. The writing is on the
wall, large and clear.*

- Jon Wynne-Tyson

*Man, do not pride yourself on your superiority to animals:
they are without sin, and you, with your greatness, defile
the earth by your appearance on it, and leave the traces of
your foulness after you.*

- Fyodor Dostoevsky.

"Rangers are the guardians of our planet's most precious natural assets and it's unnerving to think that every day they go to work, their lives are at risk as a result of human greed and cruelty. Without solid protection, proper law enforcement, and a strong support network for those unsung heroes of conservation, our efforts to protect wildlife are a lost cause," Julia Marton-Lefevre, IUCN [31].

Park rangers patrol the wildlife reserves in Africa, India, and other locations trying to save endangered animals from poachers. They fight criminal organizations that are armed with high power rifles, helicopters, and night-vision goggles while they have almost no resources, at best nothing but outdated rifles, and local laws require them to shoot only to defend themselves, giving the poachers the edge. More than a hundred of these frontline warriors die every year. Their families left behind are left suffering in dire straits, no income, no husband, and no father. Yet these heroes are on the front line against terrorism and provide a service to the world in fighting terrorist organizations that finance much of their operations from the wildlife trade and poaching [32].

It may seem paradoxical to include this discussion of park rangers in a treatise on "helping animals is helping people." A critic might say that this is a case where helping animals is *harming* people, if rangers shoot poachers who are in the act of killing animals. But the issue is analogous to the role of police forces ubiquitous in every country on Earth. Not only are some criminal perpetrators shot in carrying out their crimes but many police officers are killed in the line of duty as well. This is a tragic fact that accompanies the enormous benefit that the police are providing for the greater public welfare. The same is true of the wildlife park rangers who fight to protect what is humanity's precious inheritance - the magnificent wildlife the earth was blessed with. These rangers help to ensure that these threatened and endangered animals will still exist for posterity, exist for countless humans to experience and enjoy in future times. When they help these animals by saving their lives, fighting these poachers, they are helping countless people to come. They are ensuring that future generations will inherit a world in which tigers, elephants, rhinos, gorillas, and lions hopefully still exist.

The poaching issue represents in its essence a little-known war in which incredibly brave young men, and sometimes women, are being killed alongside the animals they are trying to protect. In the past 10 years, more than 1000 rangers have been killed in the line of duty, rangers with outmoded and outdated weapons pitted against sophisticated enemies with much higher firepower [33-34]. Rangers face death, injury, and torture from poachers, who are sometimes protected by corrupt government officials [35-36]. Penalties for poachers who are caught have historically been close to non-existent, "a slap on the wrist," a small fine, and poaching has historically been considered a low risk crime.

The problem with poaching probably started with the bushmeat trade described earlier where animals are slaughtered to provide meat for workers and to sell to markets in surrounding cities. The bushmeat trade has risen to millions of tons of antelope, elephants, and especially the great apes - chimpanzees, gorillas, bonobos - and species of monkeys. Organized crime jumped in quickly to benefit from the large cash flow, and brought "professional" killing methods with them including snares, poisons, and automatic weapons. At the same time, terrorist organizations recognized the lucrative opportunities and began joining in the slaughter. There was too much money to be made and almost no opposition, just the few and far between park rangers armed with old rifles, and governments that often sided with the poachers.

These terrorist organizations fight government forces for political reasons, kill civilians, turn children into killers and suicide bombers, and commit atrocities, sometimes destroying whole villages in rape, torture, and murder. They fund these operations partly by extortion and taxation of victim populations and partly by large scale poaching of animals. Many of the groups are Islamic jihadists eager to impose sharia law on whole populations.

Al-Shabaab is an Al-Qaeda linked group in Somalia which raises an estimated $600,000 a month by poaching [33, 37]. This is the group that killed 67 people in the Westgate Mall in Nairobi [38] and another 48 near a coastal resort. In one estimate, elephant slaughter and illegal ivory fund as much as 40% of their operations [39]. They are responsible for suicide bombings in Kenya, Somalia, Uganda, attacks on tourists, and attacks on United Nations compounds and other peacekeepers, as well as attacks on soldiers. They have recruited some Americans into their ranks. A second jihadist group, Jahba East Africa, has also begun operations in Somalia and parts of Africa, after declaring allegiance to ISIS [40].

Another militant Islamist group profiting from the ivory trade is Boko Haram, operating mostly in Nigeria but spilling over into Chad, Cameroon, and Niger [33, 41]. The group has declared loyalty to ISIS and wants to establish sharia law across Nigeria and expand it to other countries. Boko Haram has bombed schools, churches, and mosques and assassinated politicians, but it mostly targets civilians, burning men, women, and children alive in its attacks and kidnapping children to become suicide bombers [41-44]. It was responsible for over 6000 deaths in 2015 alone and as many as 20,000 overall, displacing more than 2 million refugees trying to escape the violence, and causing widespread starvation [45].

The Janjaweed is another Islamist militant group operating in the Sudan and is allied with forces of the Sudanese government, unleashing genocide which claimed 400,000 lives and displaced 2.5 million people [46]. More than a hundred people continue to die each day (year 2015-2016 time frame). As with the other jihadist groups, the Janjaweed targets civilians, mutilating and killing the men, raping the women, and kidnapping the children, as well as burning villages and fields and poisoning wells [46-47]. The Sudanese President was indicted by the International Criminal Court for the mass killing and genocide.

Perhaps the only non-Islamist terrorist group operating in Africa is the Lord's Resistance Army led by Joseph Kony in northern Uganda, parts of the Congo, Central African Republic, and Sudan. The group started as a Christian movement with the aim of ruling Uganda by the Ten Commandments (the absurdity of using the Commandments, one of which is Thou Shalt Not Kill, to justify mutilating and killing thousands of people boggles the mind). In any case, their tactics are the same as the other African terrorist groups, targeting civilians, torturing and killing opponents, kidnapping children and forcing them to become child soldiers [48-49]. LRA's activities have claimed at least 100,000 lives and displaced

2.5 million people [49]. Like the other terrorist groups, the LRA is funded largely by the ivory trade [33 37-38, 49-50].

As other terrorist groups spring up, or established ones like ISIS (ISIL, DAESH) expand into new territories, they are likely to fund part of their operations from the wildlife trade, as long as they can find a willing market both in the West and especially in Asia. From the present indifference that people exhibit toward wildlife, finding such markets probably won't be difficult.

From bushmeat to genocide to kidnapping to raping, suicide bombings, and child soldiers, and to the destabilizations of sovereign governments, the poaching of elephants, rhinos, gorillas, chimpanzees, and other African wildlife is helping to fund terrorist organizations responsible for countless murders of innocent people. "Blood ivory is now a major driver of insecurity across Africa and not only threatens elephants but also directly leads to killings of people," Kasper Agger, field researcher, the Enough Project. Park rangers are on the front line against these terrorists and are responsible for actually saving some individual populations of mountain gorillas and lions, but they are too few and far between to tackle the enormity of the battle, though thousands have died trying. It has been said that fighting poaching and the wildlife trade is a new front in the fight against terrorism [38]. Certainly the high tech tools of western military forces: drones, helicopter gunships, special forces, high powered weaponry, would make an enormous difference in stopping these trades, prevent extinction of iconic species, and save many human beings as well as many endangered animals. Fighting poaching in Africa and elsewhere is a clear example where helping animals is helping people.

TRAPPING

Oh, never a brute in the forest and never a snake in the fen
Or ravening bird, starvation stirred,
has hunted its prey like men
For hunger and fear and passion alone drive beasts to slay
But wonderful man, the crown of the plan,
tortures and kills for play.

- Ella Wheeler Wilcox

It seems fitting to end this topic of hunting with the related practice of trapping. There are leghold traps, body traps called Conibear, and snares made from rope or more typically wire. Instead of hunting with rifle or arrow, individuals "hunt" by hiding these devices where wildlife is likely to pass. They are indiscriminate, catching anything that comes their way, from targeted animals to dogs and cats, non-target wildlife considered nuisances by the disgusted trapper, and occasionally endangered species like America's symbol - the Bald Eagle.

These traps are barbaric devices that cause unimaginable terror, pain, and suffering. It's not unusual for an animal caught by the arm or leg to literally gnaw off the body part trying to escape the fear and pain. It can be said that human beings' willingness to cause this degree of suffering is an indictment, an indication of the lack of empathy and compassion in the individual's character, and the fact that governments in the form of wildlife agencies not only allow this practice but actually encourage it doesn't speak well for the human species. Recall that trapping is allowed in more than half of official Wildlife Refuges in the United States. Some refuge!

Perhaps ironically, trapping was actually begun in Europe to catch *people*, not animals. Aristocrats, particularly in England, considered the forests to be their private properties and all the wild-life in those forests belonged to them. Peasants often struggling to find enough to eat would hunt deer, rabbits, and other "game" in those forests, which didn't sit well with the elite. The lords of the manor would set large leghold traps in the forest to catch the poachers, many of which would result in broken legs, often leading to crippling or death. These instruments had alternating sharpened spikes that would impale the leg in multiple places, and many had additional vertical spikes that would impale the foot from below at the same time. They were intended to maim and punish the au-dacious interloper who would dare to enter the aristocrat's private game preserve. Eventually the savagery and cruelty to humans of the spiked traps caused such an outcry that a "humane" man trap was substituted in the 1700's. The traps were so powerful that it would take two grown men to open them; they were 5 to 6 feet long and several feet in diameter. A slightly smaller version is used today in the U.S. as "bear traps."

Another tool used to catch the unlucky trespasser in the forest was the spring gun that would fire a spray of shot into the individual when the trip wire was triggered.

Mantraps, including spiked traps, were used in World War I and even more recently in jungle warfare in the Far East. The WWI version was a rectangular box 2 feet long with interlocking 6 inch spikes, buried just below the surface. When a young soldier would step on the wrong place, the spikes would be driven into his leg from both sides while more spikes would drive upward into his foot. The rows of spikes were pointed downward, making it almost impos-sible to free the individual and leaving him to die slowly in agony. In jungle warfare, mantraps made with sharpened wooden spikes would be hidden in the brush waiting to impale the soldier, child, peasant, or animal. Contamination was sometimes added to the spikes to cause blood poisoning which would manifest days later, assuming the person survived the original impalement.

What do animals experience when caught in these leghold traps? An Englishman named Ned Niall relates his experience when caught in his own trap, a small one set for small animals like rabbits or hedgehogs. "I thrust my hand into the hole [where the trap was set] which had been empty [of a trapped animal] and it clamped firmly onto my wrist. The teeth, as sharp as those of a farm dog, buried themselves in my flesh. In agony I withdrew my hand and the trap with it. I had not the strength to remove it, for even a grown man must step on it to release the jaws. Little trickles of blood ran down to my finger tips and splashed onto my knees. I had an awful vision of myself starving to death, tethered in the corner by the wall like a goat. I lay at dusk, weak and shivering at the ghostly moan of the wind through the stones. I was more frightened than I had ever been before. When I recovered my composure I set my teeth and began the painful process of moving my wrist in the trap. Every small movement brought fresh tears. As the teeth of the trap dragged over my flesh the skin was ploughed up. Several times I had to stop and close my eyes until, finally, with the last rally of my courage, I ripped my hand from its grip. The sight of my wrist, bleeding and swollen, frightened me so that I could not look at it. I stood up and made my way through the gorse, leaving the trap, where it may lie to this day for all I know. I never set another, and once shot a fox I found tied down by one in the heart of a wood" [51]. Mr. Niall later told others how he had been in continual agony and had a badly mangled arm and fingers for years.

Interestingly, the steel trap was first used for animals in North America around 1823, about the same time that man traps, even the so-called humane ones, were outlawed as excessively barbaric and cruel in England [52].

So how is helping animals in this trapping issue helping people? Aside from the children who have lost fingers in hidden animal traps, aside from families who have watched in horror as their beloved pet is caught and killed in a trap, aside from the damage to a child who steps on one while playing in the woods, aside from allaying the fear of people trying to enjoy wildlife refuges without being shot or trapped, the benefits to people are more indirect. I'd like to think that the human character has, or will some day, evolve to where people will be unwilling to cause the unimaginable pain and suffering that animals endure when caught in these leghold traps, when humankind will develop the empathy, compassion, and understanding to allow the fellow creatures of the earth to live out their lives in peace. It's a long shot, I know, but I can still dream. If our better treatment of animals will have hastened human evolution toward kindness, empathy, and peace toward his fellow man as well as brother animals, then helping animals will have helped people indeed.

CHAPTER 3
Marine Wildlife

The education of the heart should ever go hand in hand with the cultivation of the mind. Kindness toward all sentient creatures and compassion for suffering in all its forms are the hallmarks of the enlightened community and the badge of the cultural individual.

- George Farnum

Of all the wildlife on Earth, none are as exploited, as devastated, as all the creatures living in the seas, the crabs, the lobsters, the octopus, the bivalves (oysters, scallops), but most of all the fish. Fish receive virtually no sympathy, no respect, no compassion. The millions of anglers who catch them with barbed hooks through their mouths or other parts of their body and let them die a slow death of suffocation in a pail or on a string as they struggle to breathe with a physiology designed to extract oxygen from water, not air, never give a thought to the suffering their prey is undergoing. (Anyone with acute asthma or COPD can imagine better than the rest of us what it's like to struggle for every breath, akin to drowning in a sea of air). Books like Tom Sawyer, Huckleberry Finn and many paintings of Norman Rockwell depict happy, smiling, laughing young people walking along with strings of fish over their shoulder or in a pail. But like any living creatures torn violently out of their environment, the fish experience terror and pain.

Commercial fishing operations kill fish by the hundreds of thousands, catching them on hooks on "long lines" a mile long or dragging them from the ocean depths in massive nets to dump them onto a trawler's deck. If they don't die then of suffocation, they do by being crushed by the weight of thousands of other fish above them, or by ruptured swim bladders as they are brought up from the larger pressures of the water depths to the much smaller pressure at the surface where the difference makes their bodies nearly explode [1-2]. Fish suffer a form of the bends as their bodies react to the sudden change in pressure. As in humans, the bends are associated with a lot of pain.

The commercial fishing industry is inherently cruel. Besides hooking fish on long lines baited with squid or other fish, where the fish that is caught may linger on the line many hours before brought onto the ship, many unwanted beings are unintentionally caught, including sea birds like albatrosses, turtles, marine mammals, and commercially useless species [2]. This collateral damage is known as bycatch and can be a significant part of the catch. Shark finning where the fin is sliced off for sale in the Asian market and the shark thrown back to die, bluefin tuna stabbed by piercing blunt hooks in almost any part of their body [3], catfish extreme cruelty exposed in the aquaculture industry [4-5], and the crushing, hooking, bladder-exploding consequences of trawler net fishing, all involve intense suffering. All this suffering is ignored by the industry and by society as inconsequential.

Why is it that we care so much about dogs, cats, and fuzzy mammals, and even about tigers, elephants, and lions (remember Cecil?), but don't find it in our collective hearts to care about marine wild-life and particularly fish? We don't treat other animals that way. We might turn a blind eye when it comes to factory farming and the meat on our table, but most of us don't actually want those cows, pigs, and chickens to suffer on the way to our stomachs. Charlotte's Web, the movie Babe, happy cows in commercials or milk cartons, may have even made us sympathetic toward some farm animals and feeling a little guilty (at least some of us) about the way they are treated and our part in that. We also seem to have some sympathy for horses. Humanity literally depended on them for survival for centuries and they still are the measure of power and strength in our motors and engines, yet we let them die by the thousands at race tracks and suffer terror and pain on their way to be slaughtered.

But we harbor no sympathy at all for fish. Once they are re-moved from the hook and put on the string or tossed into the bucket with others, they're left to suffocate slowly while the angler goes back for more. On board the fishing trawler, they are just "things," a means to a making a living for the fishermen, not living beings undergoing intense pain, fear, and suffering (scientific experiments show that fish do experience pain and fear [1] - more on this later).

It has been suggested that much of this indifference stems from the view that fish are so *alien* to us [1-2, 6-7], and in fact the entire ocean environment is alien to us, less understood than the surface of the moon once we venture beyond the seashore where we like to swim. Fish can't smile, can't yell and scream when in pain, can't close their eyes because they have no eyelids; they breathe by an en-tirely different mechanism than us, in fact seem to "breathe" water that would make us drown. We humans have another bias; to us, all the fish of the same type look alike - all the trout look like trout, all the sunfish look like sunfish - they're all the same. Yet fish them-selves can discriminate between individuals of their own species and between members of other species.

Bring up a hundred thousand fish in a trawler's net and they all look alike to us, yet they are actually all individuals with individual personalities [2]. To use an analogy, imagine a football stadium filled to capacity with 100,000 people. We have no problem thinking that each of the 100,000 is a unique individual with a unique personality. Carry it a step further; there are 7 billion people in the world today, yet we readily accept that each one of them is different and unique. Why is it so hard to contemplate that each of those 100,000 fish is also unique in their own way, in their own world, even though they all *look* alike? Each is a sentient individual with an instinct and wish to live like any other life form.

The issue of whether fish feel pain has long been settled by dozens of scientific studies, including their reaction to pain killers like opioids [1-2, 7-13]. The issue of sentience was also settled long ago [1-2, 11-12, 14]; in fact, fish are sentient and self-aware. Some of the 30,000 species of fish exhibit considerable intelligence [1-2, 7-8] and can learn a variety of tricks and skills, while having excellent memories [1-2, 7-9, 12, 15-16]. Fish recognize other individuals [2, 7, 12], can communicate with others [2, 8], and can plan ahead and solve problems [1-2]. Similar skills have been seen in marine invertebrates; octopuses have been known to figure out how to open twist-cap jars and child-proof bottles (something 5-year old children can't do), and have been seen to escape from their tank, go on a nightly sojourn, and return before lab technicians arrive the next morning [1, 15], which exhibits planning, purpose, and memory. Groupers and eels have been known to hunt together as a team; the grouper communicates to the eel where good tasting prey can be found but is hidden by a reef or rocks. Once the eel scares the prey out of hiding, the two share in the prize.

Fish have been observed using tools [2, 17]. In a similar way that birds open nuts by dropping the nut onto a rock, fish have opened oyster and clam shells by holding the shell in their mouth and smashing it on a rock. (Birds have it a lot easier; air is a lot less resistant than water, so fish can't just drop the clam onto the rock.) Fish have been seen to build nests using jets of water from their mouths to shape the sand, and to uncover hidden prey using jets of water [17]. They have been seen using leaves to transport their eggs to remove them from danger or move them to a new nest. Archerfish use jets of water like bullets fired from their mouth to knock down insects resting on leaves or branches above the water. This is a skill taught by one archer fish to another, perfected with practice.

Fish have even been known to play games, pushing a thermometer floating in an aquarium to make it wobble back and forth, and even playing a hide and seek, cat and mouse-like game, with a cat [2]. There is widespread belief that when fish, dolphins, and whales jump out of the water and come crashing back down with a mighty splash, they are doing it out of fun and entertainment. Fish have also been known to help other fish that are sick or injured, exhibiting sympathy and compassion [2] that most people would never believe could come from a fish.

By now after reading this we all should realize that fish are not robotic "things" but instead are individuals with many traits much the same as land animals including our beloved dogs and cats. Yet the devastation carried out on fish exceeds all other life forms combined. Globally, between 1 and 3 *trillion* wild fish and 100 billion farmed fish are caught and killed [12]. Hundreds of millions more are caught for sport. This doesn't include the by-catch, the unwanted fish, turtles, and marine mammals that are caught up in the trawler's nets and subsequently die. Once it seemed that the oceans and the sea life in them were vast and boundless. Now we know that fish populations in the oceans have been reduced to half their former number while certain species like tuna and mackerel have been reduced by 75% [18]. The populations of ocean fish are rapidly collapsing due to overfishing, coastal development with pollution and runoff, coral reef and mangrove destruction which are nurseries to many fish species, and effects associated with climate change. Populations of marine mammals, birds, and reptiles have also fallen close to 50% [18].

Corals are tiny marine animals responsible for the fabrication of the magnificent coral reefs that surround many sea coastlines. They play a vital role for mankind in helping to protect coastlines from damaging storms and in forming habitat and birthplaces for count-less fish and shellfish species [19-23]. "Coral reefs are invaluable resources to local communities around the world, serving as sources of food, jobs and livelihoods, and as coastal protection" [21] and more than 100 countries worldwide rely on them for income, food, and storm protection [19]. Yet coral reefs are in real danger from several man-made causes. One of these is destructive fishing where dynamite is set off around reefs to kill numerous fish which are then collected by local fishermen. The dynamite destroys nearly all the life forms around the reefs including the corals that built them and parts or all of the reefs themselves, ruining all the benefits that hu-mans derive from the reefs. Cyanide poison is used to stun or kill both food fish and hobby (aquarium) fish, but destroys the corals at the same time. More damage comes from global warming - climate change - that makes the oceans more acidic and raises the ocean temperatures; even a small temperature change is enough to kill many species of coral. It is estimated that 25% of the world's coral reefs have already been destroyed because of climate change [24] and one-third of corals face extinction from the combined effects of climate change and the other man-made causes already mentioned [25-27], including pollution and agricultutal runoff.

What about diseases? It might be surprising to hear that just as there are many diseases that afflict land animals and humans, there are also many diseases that afflict sea life as well. Even though sea water can act as something of an antiseptic for minor external human infections, the sea water is host to its own bacteria and other pathogens. Over the years there have been many epizootics, the word for epidemics in non-humans species, diseases that have destroyed fish, shellfish, bivalves, corals, and marine mammals [28]. There are various species of vibrio bacteria that affect humans and marine life alike. One of these, vibrio cholerae, has killed millions of human beings and can be present in contaminated water, while some species of fish have been found to be reservoirs of the bacteria [29]. Vibrio vulnificus and vibrio parahaemolyticus are bacteria found in sea water that can cause human sickness and death. Open wound exposure to contaminated sea water and eating seafood containing vibrio infections are routes to human infection [28]. These vibrio bacteria also affect marine mammals, as do other serious infections like toxoplasma gondii, which is a serious infection in humans, and various forms of distemper viruses, which cause die-offs of dolphins and seals.

In some cases fish, shellfish, and bivalves are carriers of the pathogen but not afflicted themselves; in other cases the marine wildlife may contract and suffer from the disease. Human infection may come from eating the seafood, from contaminated water, and from handling and preparing the item for consumption. The list of potential pathogens is long and disturbing [29-31] and includes mycobacterium, streptococcus, photobacterium damselae, various species of vibrio bacteria including cholera, listeria, clostridium botulinum, shigella, salmonella, and campylobacter. Fortunately, cooking at high heat or sometimes freezing kills nearly all of these pathogens, which is why disease epidemics in humans caused by these bacteria are rare. Raw fish and shellfish on the other hand do carry some risk, as the pathogen, if present, may be alive and virulent.

Infectious diseases in oysters have been studied extensively due to their high commercial value. Dermo disease, a bacterial infection, has caused extensive oyster mortality in the southern U.S. coastal areas and has been moving northward with climate change. The same is true for outbreaks of MSX disease caused by a parasite, also spreading northward. Massive fish kills have taken place due to herpesvirus disease, viral hemorrhagic septicemia, and ichthyophonus disease caused by a protozoan species with characteristics of both animals and fungi. Toxic algae (algal blooms) kills many species of fish, dolphins, and other marine life, and can result in sickness and death in humans who consume the contaminated fish.

Other marine wildlife, the cephalopods (octopus, squid, cuttlefish, and nautilus) and decapod crustaceans (lobster, crab, and crayfish) have become food sources for many people, but there is very little information on possible health issues. These animals are perhaps even more alien-like to humans than fish: multiple arms, huge claws, long antennae. What they share in common with fish is the ability to feel pain and to suffer [15, 32]. It has been mentioned that octopuses are highly intelligent, solve problems, use tools, learn easily and have good memories [15]. Lobsters can live to be 100 years old, have highly developed senses of smell and taste, and have complex nervous systems. There are methods that have been developed to anaesthetize, stun, or euthanize these sensitive beings prior to cooking them [15], but people subject them to being boiled alive instead for convenience, while turning a blind eye to, or even denying, the pain and suffering this causes.

Lobsters suffer from a number of diseases, including bacterial infections, parasites, bacterial shell disease, and fungal infections [33-34]. For humans, eating undercooked lobster poses the same danger as eating any other type of undercooked shellfish, including vibrio bacterial infections, hepatitis, parasites such as roundworm and tapeworm, and toxins from toxic algae [35], though the lobster shell disease itself is apparently not dangerous to people. Eating *live* octopus is common in South Korea and Japan, but the tortured creature gets his revenge when six people a year die from choking on the tentacles [36].

Although I'm not naïve enough to think that the world will cut back on eating fish, shellfish, lobster, or the rest, at least they should do so knowing the possible consequences to themselves and to their food sources. It seems rather clear that helping marine animals by not killing or otherwise destroying them and, equally important, not damaging or contaminating their habitat, could save many people. Of course, critics would say that many millions of people get their protein from seafood while millions make their livelihood from the sea, but when the fish populations have been effectively destroyed and aquaculture can't safely take up the slack, people will need to find other jobs while millions will find other sources of protein than seafood. From the various diseases that people acquire from eating sea food (more on this will be described later), to not depleting the oceans of its living beings, to not destroying reefs that many people depend on, to not dynamiting or poisoning whole ecosystems, to not destroying whales, dolphins, sea birds, and sea turtles, to not driving species to extinction, to not (conveniently) treating fish as robotic "things" that excuses us humans for their appalling treatment, humans could benefit themselves as well as the oceans by treating marine life better. All these many species are vital for the healthy ecosystems of the sea, the health of the ocean environment. Even the tiny plankton that are the base of the marine food chain are vital to us as the main source of the earth's oxygen. When we damage the ocean by dumping our garbage, our square miles of plastic waste, our thrown-away fishing nets, our pollution, our oil spills, our overfishing, our extinction of species, our ocean acidification and temperature rise, we damage ourselves as well.

CHAPTER 4
Ocean Wildlife, Human Health, and Human Risk

*When we are confronted with the astronomical numbers
of animals who fall victim to our ocean plundering, we
struggle to make an emotional connection to them. If
there is one overarching conclusion we can draw from the
current science on fishes, it is this: fishes are not merely
alive - they have lives. They are not just things, but beings.
A fish is an individual with a personality and relationships.
He or she can plan and learn, perceive and innovate,
soothe and scheme, experience moments of pleasure,
fear, playfulness, pain, and – I suspect – joy. If we were to
interact with …any of the anonymous fishes hauled up to
their deaths, we would come to know them as individuals.
They would become someones, not somethings.*

- Jonathan Balcombe

The health benefits of eating seafood have been advertised many
times. The U.S. government, the National Academy of Sciences,

private organizations, the American Medical Association, and numerous universities across the country have lauded the benefits of including seafood in the diet [1-4]. Seafood, particularly fish, is said to be a good source of protein, low in saturated fat, and high in nutrients including selenium. It is a source of several valuable omega-3 fatty acids, such as EPA and DHA, which are also found in other sources such as several vegetables, nuts, and seeds, but in higher quantity in seafood. Eating fish has been described as reducing cardiovascular risk, partly because of the EPA and DHA, and possibly because substituting seafood for land-based meat is said to reduce the intake of cholesterol, saturated fat, and triglycerides [2, 4]. Other benefit claims of eating seafood include improvements in cognitive and visual development in infants and children, reduced risk of heart attack, stroke, and hypertension, and that fish are a good source of B-complex vitamins, vitamin D and vitamin A.

However, somewhat contrary descriptions from the National Academy of Sciences have disputed some of these claims. "Evidence for a benefit associated with seafood consumption or fish-oil supplements on blood pressure, stroke, cancer, asthma, type II diabetes, or Alzheimer's disease is inconclusive. Whereas observational studies have suggested a protective role of EPA/DHA for each of these diseases, supportive evidence from randomized clinical trials is either nonexistent or inconclusive." --- "evidence is inconsistent for protection against further cardiovascular events in individuals with a history of myocardial infarction from consumption of EPA/DHA-containing seafood or fish-oil supplements" [2].

Whether individuals choose to include seafood in their diets is a matter of personal choice, of course, but at least they should make that choice with all the facts at hand. Needless to say, helping marine wildlife means not killing and eating them. If consuming seafood has any harmful effects on people, then helping marine life in this way helps people as well. A look at the references cited above and many others may give a clue to a hidden problem. Most of these references contain warnings either in their titles or in their text; for example, "Seafood Choices, Balancing Risks and Benefits" [2], and "Fish Intake, Contaminants, and Human Health: Evaluating the Risks and Benefits" [1], and still more, "Researchers Find Major Gaps in Understanding Risks, Benefits of Eating Fish" [3]. The problem is that our oceans and the ocean wildlife that lives in them are contaminated in many ways: viruses, bacteria, parasites, chemicals, toxins, and pollutants [2, 5-7]. One might hope that the cooking is sufficient to kill the bacteria and parasites, but cooking does nothing for most viruses and chemicals, while sushi and shashimi generally are made with raw fish and don't employ cooking at all.

A common virus picked up from eating fish and shellfish is Norovirus, a group of viruses that can cause fevers, headaches, nausea, stomach pain, fatigue, and diarrhea [6-7]. Most outbreaks take place in closed, dense populations such as child care centers, cruise ships, long-term care facilities, or hospitals. Norovirus is a leading cause of food-borne illness and outbreaks. A second virus that can cause debilitating symptoms is Hepatitis A, which is the most serious viral infection associated with seafood consumption. "Most commonly, it is an acute, self-limited illness associated with fever, malaise, jaundice, anorexia, and nausea; symptoms may last from several weeks to several months. However, deaths from fulminant hepatitis can occur" [7].

The number of virus pathogens that can come from eating seafood pales in comparison to the number of bacterial species that can cause food poisoning. Fortunately most infectious cases only cause temporary symptoms but some deaths have also occurred [7]. The list of bacterial agents, mentioned earlier, includes salmonella of several varieties including typhoid, campylobacter, listeria, shigella, clostridium botulinum, escherichia coli, vibrio cholerae and other vibrio species, giardia, and staphylococcus aureus. Symptoms include combinations of stomach pain, nausea, diarrhea, dizziness, cramps, vomiting, headache, fatigue, and muscle/joint pains. Much of these pathogens come from sewage-contaminated areas and could be greatly reduced if more hygienic infrastructures were used.

Parasites are yet another fish contamination. Many fish contain species of fish tapeworms, flukes, and roundworms [2, 4-5]. Many worms are present naturally in marine and fresh waters, and most marine animals are infected. The parasite problem may be even worse in farmed fish than wild-caught fish [8]. Freezing or sufficient cooking kills these organisms, however, and sometimes they can be hand-removed when large enough to be visible. They are a bigger problem when fish is eaten raw.

Perhaps the most dangerous seafood poisoning comes from toxins sometimes present in fish and shellfish. The worst of these is ciguatera poisoning from ciguatoxin [2, 6-7, 9, 13]. This toxin comes from fish that have eaten seaweed or harmful algae containing plankton called dinoflagellates that create the toxin, or eat smaller fish already contaminated; the toxin bioaccumulates up the food chain. This is also one of the poisons that can come from the so-called toxic Harmful Algal Blooms, which have increased in recent years in response to climate change and increased agricultural runoff [10]. Ciguatera illness can last for weeks or even months, with symptoms that can include nausea, diarrhea, numbness, dizziness, headaches, slow heartbeat, high blood pressure, chills, joint pains, and mental confusion. Ciguatera can be debilitating, with severe neurologic symptoms, and may recur throughout a person's life. It can't be cured and may result in long term disability, but most people recover after a long time [6]. The best defense is to avoid it in the first place, avoid eating seafood coming from tropical waters [6].

Scombroid poisoning is a second type of food poisoning that comes from eating fish that hasn't been properly refrigerated or has been handled unhygienically [6]. Decaying fish releases a toxic histamine that can cause nausea, diarrhea, stomach cramps, headaches, rapid heartbeat, and in severe cases swollen tongue, blurred vision, and difficulty in breathing. This is a far less serious food poisoning than ciguatera and may only last 24 hours, and an antihistamine can bring some relief.

Added to these somewhat naturally-occurring dangers of consuming seafood described above are the myriad of man-made chemicals, pollutants, and pesticides that may contaminate both fresh water and sea water. Many of these are neurotoxins and some are carcinogenic. One of the worst offenders is methylmercury (CH_4Hg), a form of the metal mercury converted into the methyl compound by microorganisms in the water, while the mercury itself comes from the air. Historically, coal-fired power plants have been a major source of this air-borne pollutant, though some may also come from volcanic eruptions. CH_4Hg is a neurotoxin that can penetrate into the brain, and is of particular worry for pregnant women because it may harm the development of the fetal brain and possibly the heart and circulatory system [1-2, 5, 11-13]. It may also be a possible carcinogen [11]. Methylmercury bioaccumulates in the food chain as bigger fish eat smaller ones. The greatest concentrations are found in the major predators: swordfish, marlin, large tuna, pike, barracuda, shark, mackerel, and walleye, as well as seals and toothed whales such as the pilot whales consumed in the Faroe Islands. As prevention, the U.S. Govt. suggests that people eat fish low in methylmercury [12].

Persistent Organic Pollutants (POPS) are another class of chemicals that contaminate the oceans and can accumulate in seafood. Dioxins are a group of these pollutants that are highly toxic and can cause reproductive and developmental problems, liver damage, and an impaired immune system [1-2, 14-15]. It is especially dangerous to fetuses and newborns. Dioxins come mainly as industrial waste products and to a much lesser extent from volcanic eruptions and forest fires, but also from incomplete burning of solid waste and hospital waste. Dioxins are found throughout the world. Most human exposure comes from food - mainly meat and dairy products, fish, and shellfish.

PCB (PolyChlorinated Biphenyl) is another widespread organic pollutant found in seafood: both freshwater and saltwater fish and shellfish. It is another industrial product, used extensively in electrical machinery in the early to mid 1900's. Production was banned in 1979 but it persists in the environment, especially in rivers and other bodies of water where it was sometimes dumped as a waste product. Like the dioxins, PCBs become stored in the fat tissue and aren't excreted, so they bioaccumulate in the food chain [14, 16]. Like dioxins, PCBs are classified as probable human carcinogens [2, 16] by several government research agencies including the EPA, the National Toxicology Program, and NIOSH; "studies in humans have found increased rates of melanomas, liver cancer, gall bladder cancer, biliary tract cancer, gastrointestinal tract cancer, and brain cancer, and it may be linked to breast cancer" [16]. Other adverse effects include damage to the immune, reproductive, and neurological systems [2]. Presumably these illnesses are dose dependent, and the less PCBs (and dioxins) in the body, the better.

Other persistent pollutants found in fish and other seafood include polyaromatic hydrocarbons, chlordane, heavy metals like cadmium and lead in addition to mercury, and the long-banned pesticide DDT [2].

Many studies have been done on the health effects of all these contaminants, and many of the adverse health effects have been already described. Adverse effects are very much dependent on the quantity of the contaminant, usually measured in parts per million since the amounts present are usually and fortunately very low. A glaring omission in these studies is the question of what happens when several of these are present at the same time, perhaps multiple

pollutants but possibly infectious agents as well. Chemical pollutants can depress the immune system, making the body more susceptible to bacteria, viruses, and toxins. There are many discussions and
warnings about human health effects of pollutants and pathogens
coming from private and government organizations [1-9, 11-21].

Seafood has been described as healthful and an important food
source for many millions of people, but even the claimed health
benefits of fish and fish oil are not without their detractors [22-23].
"A systematic review and meta-analysis published in the Journal of
the American Medical Association looked at all the best randomized
clinical trials evaluating the effects of omega-3 fats on life span,
cardiac death, sudden death, heart attack, and stroke. These included
not only fish oil supplements but also studies on the effects of advis-
ing people to eat more oily fish. Overall, the researchers found no
protective benefit for overall mortality, heart disease mortality, sudden cardiac death, heart attack, or stroke" [22, pg. 34]. Consumers
should be aware of the benefits and risks when making the choice to
eat seafood, as has been pointed out by the U.S. government many
times. The possible negative effects of eating seafood are seldom
mentioned along with the advertised benefits.

The problem of contamination is even greater for the ever-increasing factory-farmed fish (aquaculture) which can contain many
of the same pollutants, but in addition antibiotics and drug-resistant
bacteria, dyes to improve color, fungicides to prevent fungus growth,
and pesticides [2, 5, 8- 9, 14, 24-25]. "A 2004 study in the journal
Science found that farmed Atlantic salmon have such high levels of
PCBs that the authors advised against eating more than one meal
per month of farmed salmon from certain areas based on the EPA's
recommended exposure levels. Although all fish accumulate POPs
from the environment into their fats, farmed fish do so at a higher
rate than wild fish" [14]. Aquaculture is also bad for wild fish, as
farmed fish sometimes escape from confinement and contaminate

the gene pool of their wild cousins, while feeding farmed fish requires large quantities of wild-caught prey fish as feedstock, which can cause overfishing of wild fish.

Climate change makes most contaminant problems worse. Warmer air and seawater temperatures, increased acidity, rising sea levels, changes in oxygen content, reduced upwelling of nutrients, all contribute to making bacterial and toxin problems greater [5, 13]. Warmer temperatures increase the occurrence and size of Harmful Algal Blooms which expand ciguatoxin production, while populations of vibrio species [13, 26-27] and other harmful bacteria are enhanced by warmer temperatures. Coral bleaching provides more surface area for toxin-producing marine algae and dinoflagellates to grow. Toxic heavy metals like cadmium, lead, and mercury are expected to increase as reduced salinity near coasts increase the uptake of these metals by phytoplankton which are then eaten by fish. The geographic range of pathogens is also increased as warmer temperatures expand northward and southward, contaminating a larger fraction of fish species. Fish and marine mammals suffer from the increased effects of climate change as do humans [26-27], and helping marine wildlife helps people at the same time.

CHAPTER 5
Climate Change

It took 10 years to get to the point where it was accepted that there were NOT two equally valid sides to climate science. Millions of dollars were spent by people deliberately trying to confuse that issue, and they are doing it again today. They're doing it on the cost issue because they lost the science battle. They're trying to scare the public into thinking this bill is going to put people out of work and damage the economy.

- Peter Goldmark

TIMELINE.

1965. Climate scientists warn political leaders that the greenhouse effect could become a real problem.

1974. Physicists point out that the possible severity of GW might require humanity to leave some fraction of fossil fuel reserves in the ground. (GW = global warming).

1977. The National Research Council warns about a serious impact of unimpeded climate change.

1978-80. The National Climate Act establishes a national climate research program in the U.S. The National Academy of Sciences forms a study committee. The 1st report doesn't deny the climate science but specifies that it would be too expensive to fight the causes, that new technology would come along, and people could just migrate.

1988. The Intergovernmental Panel on Climate Change (IPCC) is established by the United Nations Environment Program and the World Meteorological Organization in recognition of the seriousness of global warming.

1990. The IPCC publishes its first report on climate change. One conclusion is that unrestricted use of fossil fuels would result in a mean temperature increase of about 0.3 degrees C (0.54 degrees F) per decade, a number which has been largely born out.

1990. Warnings of the possible consequences of global warming are given: drought and famine, floods, rising sea levels, submerged coastal areas.

1992. Science reports appear about rising greenhouse gases in the atmosphere: CO_2, CFCs, CH_4, N_2O, H_2O. More warnings are given - higher temperatures, increased storms, damaged natural ecosystems.

1992. Climate change treaty meeting (Earth Summit) is held in Rio de Janeiro, including 108 heads of state and representatives of 166 nations. The developed nations pledge to reduce emissions of greenhouse gases by the year 2000.

1994. The governments of 192 countries sign onto the Framework Convention, pledged to help prevent dangerous anthropogenic interference in the climate system by reducing greenhouse emissions.

1995. Second report of the IPCC, stating that GW is at least partly man-made and could lead to more intense storms, destruction due to rising sea levels, diminished agricultural output, increase in diseases, and ecosystem damage to systems that can't adapt quickly enough. Global average temperature rise is noted to have been 1 degree F since start of the industrial age.

1996. Madrid meeting of climate scientists report that GW is real and partly man-made. Strong opposition arises from political sources and several well-known scientists (Seitz, Singer, Lindzen). Conservatives in the U.S. Congress strongly oppose the Madrid report's conclusions.

1997. Wall Street Journal editorializes that GW is a myth, citing satellite and balloon data that show no temperature increase in the atmosphere and also suggests that we expand and unregulate fossil fuel burning.

1997. Global climate meeting takes place in Kyoto with more than 160 countries, 5000 delegates, to discuss how to combat climate change. Industrial nations agree to cut emissions by 2012 by 5% below 1990 levels; China and India are exempted. The cuts are believed to be way too small to make much difference. CO_2 (carbon dioxide) levels reach 360 ppm (parts per million) compared to pre-industrial levels of 280 ppm.

1997. Letter is sent to world leaders from hundreds of medical professionals warning of health risks of climate change.

1998. U.S. signs a treaty in Buenos Aires to limit greenhouse gases but U.S. Senate opposes it and opposes the Kyoto protocol. Opponents say it would be too costly to the economy to fight GW. American Petroleum Institute is said to spend millions to recruit scientists who will oppose GW and train them in public relations to convince the public that climate change isn't real.

1998. The widespread assumption that the effects of climate change will be felt gradually and smoothly, allowing adaptation to the changes, is seriously questioned, and opinions increase that if the climate changes, it may change quickly instead, within a human lifetime or even less, a few decades.

2000. Climate meeting at The Hague, Netherlands, draws protesters saying progress on fighting climate change is too slow.

2001. The IPCC publishes its Third Assessment Report on climate change in several parts - "The Scientific Basis;" "Impacts, Adaptation, and Vulnerability;" and "Mitigation." A U.N. conference on climate change is held in Shanghai, attended by 99 nations. Their report emphasizes that climate change is caused by man-made pollution, not changes in the sun or volcanic eruptions. The report warns of increased droughts, melting glaciers and sea level rise, and flooding.

2002. The Larsen B ice shelf, the size of Rhode Island, 1250 sq. miles, which was locked to the shore since the last ice age,

abruptly disintegrates into a constellation of icebergs. Unusually warm Antarctic summers are expected as the cause.
http://earthobservatory.nasa.gov/Features/WorldOfChange/larsenb.php

2004.   A relook at satellite data and balloon data on warming in the atmosphere clears up some discrepancies and is consistent with on-going global warming. Previous errors in satellite and balloon data are resolved and prior *political* conclusions made with erroneous data are shown to be wrong. (See Wall Street Journal article in 1997)

2004.   A report by the Union of Concerned Scientists criticizes the Bush administration for manipulation, suppression, and misrepre-sentation of the science on global warming. The report is endorsed by 60 influential scientists including 20 Nobel laureates. In Buenos Aires, a United Nations conference on climate change ends with little progress after opposition from the United States.

2006.   A documentary "An Inconvenient Truth" debuts at the Sundance Film Festival showcasing former Vice President Al Gore's slide presentation and warnings about climate change.

2007.   The IPCC publishes its Fourth Assessment Report which builds upon the sections of their 2001 report. The IPCC and Al Gore share the Nobel Prize for their efforts to build knowledge of man-made climate change and formulate measures to counteract such change.

2007.   It is recognized that much of the extra heat caused by the greenhouse effect is taken up by the oceans (90%) which results in a much slower rise in atmospheric temperatures, delaying the worst effects of rising *atmospheric* temperatures and seemingly masking its effects from the public.

2010. A proposed "Cap and Trade" plan by the Obama Administration dies in Congress. Economists believe a tax on carbon emissions is the better way to fight climate change.

2012-2013. Definitive scientific studies using temperature readings, tree ring data, and sea-floor and land sediment data provide accurate information that the Earth was warmer at the start of the 21st century than at any time in the last two millennia.

2013. The IPCC publishes "Climate Change 2013: The Physical Science Basis. A Summary for Policymakers."

2014. A new record is set for global atmospheric temperature. According to the American Association for the Advancement of Science (AAAS), 97% of climate scientists have concluded that human-caused climate change is happening.

2015 The average atmospheric temperature exceeds even the 2014 record, with the aid of an El Nino, reaching a landmark 1 degree C (1.8 degree Fahrenheit).

2015. A climate change meeting in Paris includes 195 countries. It is the 1st global accord to commit nearly every nation to take domestic action to fight climate change.

2015. Obama administration regulations regarding mercury and other air-borne toxins and CO_2 emissions from the electric generation sector are formulated, requiring a 32% reduction in CO_2 emissions below the 2005 level by the year 2030. The regulations come from the EPA under the Clean Air Act.

2016. The year sets yet another new record for average global temperature, surpassing the record years of 2014 and 2015.

2016. The American Institute of Physics publishes a history of the climate change/global warming debate and shows that any significant doubt in the scientific community about its reality and man-made causes is virtually eliminated.
https://www.aip.org/history/climate/20ctrend.htm

2016. The U.S. Supreme Court blocks the Obama administration's regulations to curb greenhouse gas emissions. The Attorneys General of several states sue Exxon-Mobil stating that the company's research on climate change showed results that differed from their public announcements and that Exxon-Mobil spent years funding organizations that worked to undercut climate science. Many scientists express views that fossil fuels should be left in the ground in order to prevent catastrophic climate change, but this would reduce "Big Petroleum's" and "Big Coal's" net worth, some of which comes from in-ground fossil fuel reserves with the expectation that they will be used.

2016. President Obama rejects the Keystone Pipeline application that would carry oil from Alberta tar-sands to Gulf Coast refineries.

2016. U.S. President Obama and China President Xi Jinping agree to sign the Paris Agreement and reduce their country's emissions.

2016. Pentagon planning documents recognize climate change as a threat to National Security, possibly leading to a massive and increased flow of refugees, violent conflicts over natural resources such as food and water, and increased natural disasters.

2016. The U.S. Global Research Program publishes <u>The Impacts Of Climate Change on Human Health, A Scientific Assessment,</u> available at https://health2016.globalchange.gov/downloads

2017. Newly-elected President Trump reinstates the Keystone and Dakota Access Pipelines. The Trump administration removes all references to climate change from the White House Website and orders Govt. scientists to obtain political approval before publishing any scientific data. Scott Pruitt becomes EPA head, denies climate change exists and pushes to reduce the agency. Funds to continue the study of climate change and mitigate its effects are removed from the budget. President Trump withdraws the United States from the Paris Accord.

DYNAMICS AND CONSEQUENCES OF CLIMATE CHANGE

Outside of a nuclear war, no other global issue is as consequential and threatening to both people and animals as unhindered climate change, and no other activity is as helpful to both as mitigating and hopefully preventing it. Animals and people have adapted well to the climate that has existed for the last thousands of years. Sudden change over a few decades, with many extreme events taking place at increasing frequency, brings new difficulties for both to adapt to the new reality.

Scientifically, there is overwhelming consensus that global warming and climate change are occurring and that human activities are the primary cause. Politically, the issue remains contentious and divisive. The world leaders alive today, the heads of state and their governments, are generally old enough that they won't bear the pain and the penalty; they establish their policies for short term gain and short term political advantage. It is future generations, perhaps people in their 20's and 30's today, and certainly their progeny, that will suffer the long term consequences of this short-sighted denial of a planet-wide threatening event.

Naomi Oreskes and Erik Conway have written a remarkable book entitled Merchants of Doubt, How a Handful of Scientists Obscured the Truth on Issues from Tobacco Smoke to Global Warming [1]. What these authors show is that, just as there was a systematic attempt by proponents of tobacco smoking to obscure and deny the health aspects of smoking, there has been a similar attempt by climate change opponents and their supporters, particularly the coal and petroleum industries, to trivialize and deny climate change. In some cases, the same few scientists who denied the smoking - health connection are the same ones who denied the reality of climate change, and these few scientists sometimes selected data to bolster their case while omitting data that showed the opposite conclusion, something that is antithetical to good scientific practice.

While an overwhelming majority of climate scientists are convinced that GW-CC is real and largely anthropogenic, the fossil fuel industries, countries heavily dependent on oil sales, and members of various governments, especially the U.S. government, have used every trick they could to deny the science and convince themselves and the public that there is no validity to the concept. The misinformation fed by the handful of deniers was enhanced by the mass media that appeared to believe equal status should be given to the nay-sayers as given to the preponderance of scientific evidence and great majority of climate science experts. What can the public be expected to believe when the news media quote the handful of climate change deniers and politicians equally with the far larger group of climate scientists who have no political agenda in their research and conclusions!

The United Nations notwithstanding, the several hundred

nations of the Earth have a difficult time working together to solve common problems. Politics, tribal disputes, unequal resource distribution, religious differences, racism, all seem to work together to prevent collective cooperation and collective action. It's difficult to see how the world's nations will come together to solve a problem so monumental that it potentially threatens the present and future lives of almost all living beings. The planet as we know it will not be the same in several hundred years unless a truly major paradigm shift takes place.

Scientifically, there can be a considerable delay time in climate change between cause and effect. It takes decades for glaciers to melt, decades for permafrost to melt, decades for atmospheric temperatures to rise. Most importantly, the majority of the extra heat endangering the earth is absorbed in the vast oceans (out of sight, out of mind?) which will delay the truly serious effects for a number of years. The heat balance of the earth is very simple; "heat in" from solar radiation and man's conversion of <u>stored</u> energy into heat (that is, burning fossil fuels and releasing nuclear energy) is balanced by heat radiated from the earth back into space. This balance creates the "average temperature" of the planet. Greenhouse gases tip that balance toward retaining heat energy rather than releasing it into space, so the global temperature rises. The oceans acting as heat sinks means that the atmospheric temperature rise is delayed from the ongoing rise of the CO_2 concentration. The inertia in the system means that the progeny of today's living human beings will suffer most from that delay instead of those alive today.

Some of the forces driving climate change contain "positive feedback" phenomena, where an effect once started continues on its own. One of these is the earth's albedo - the average reflection of sunlight back into space before it can be absorbed. Snow and ice, being stark white, act almost as mirrors reflecting sunlight. As the atmosphere warms and the ice and snow melt, the sunlight is absorbed by the darker "revealed" land area below, which increases the warming and melts even more of the remaining snow and ice, causing even more sunlight absorption. In addition, water vapor is a potent greenhouse gas. As the temperature rises, there is more evaporation, and the extra greenhouse gas retains even more heat, further increasing the temperature and thus causing more evaporation, retaining even more heat, and so forth, effectively a downward spiral.

Other examples of positive feedback for global warming include the melting of floating sea ice which increases the amount of sunlight absorbed by the water which raises the water temperature, which melts more sea ice, and so on, and the melting of permafrost which puts CO_2 and methane into the air which raises the temperature which melts even more permafrost, and so on.

The earth's temperature rise is greatest in high latitudes, such as the Arctic and Antarctic, where much of the ice and snow reside, compared to the lower latitudes, causing faster melting in the far North and South than if the rising temperatures were globally uniform. These higher temperatures increase the rate at which glaciers and other land ice masses melt, which can dump large amounts of freshwater into the northern or southern oceans, disrupting the engine that drives global ocean currents [2] and potentially altering weather patterns from Africa to Greenland and South America to Canada.

The National Geographic Society has created a "Global Warming Consequences World Map" which shows many of the consequences of climate change as they impact different regions of the world [3]. A partial list of these consequences, according to NGS, includes damaged forests, loss of biodiversity, extinction of species, rising sea levels, heat waves, water shortages, reduced snowpack, reduced growing seasons, severe storms, flooding, droughts, reductions in sea ice, threatened cultures, threatened island nations, receding glaciers, increased diseases, and expanded disease ranges. To these could be added increased iceberg hazards, mudslides and landslides, wildfires, refugee crises, decreasing resources, decreases in habitable land area, and increasing human and animal deaths.

It's important to realize that these serious potential consequences don't *have* to happen. They're the result that comes about if we don't do anything to *prevent* catastrophic climate change. These predicted consequences come from two science-based sources: sophisticated, very complex, and increasingly accurate mathematical models used by climate scientists, and extrapolation of climate events being seen all over the world today: increasing heat waves, more violent storms, rising sea levels, and so on. Similar mathematical models are used in every day society, to study traffic patterns, life insurance, actuarial tables, agriculture, disaster planning, economics, medicine, and many others. It is because of these extraordinary climate models and the ever-increasing wealth of data that goes into them that we can understand the potential consequences of our world-wide actions and have a chance of altering the path that we're on.

Temperature.

When people think of global warming, they usually think about the temperature they experience outdoors in the summer or winter. Even with the oceans acting as a massive heat sink, average *atmospheric* global temperatures have been rising year after year. At the end of the last century, it was reported that "7 of the 10 warmest years have occurred since 1990" [4], and further reported that "global temperatures in 1998 shattered the record high mark, making it the warmest since at least 1860, and possibly since the end of the last millennium" [5]. The Pacific Ocean cyclical event known as El Nino contributed to the warming effect and to droughts in some places and heavy rains in others [6]. Paradoxically, climate warming in one location can cause cooling in another, particularly the

Northern Hemisphere, by interfering with the massive ocean current that normally conveys heat from lower latitudes up the coast of Africa and Europe [7-8].

Subsequent years continued the rising temperature trend. The decade between 2000 and 2010 saw record temperatures nearly every year, and record temperatures continued in the years since. "Global Temperatures Highest in 4000 Years" [9]; "2015 Far Eclipsed 2014 As World's Hottest Year, Climate Scientists Say" [10]; "For the Third Year, The Earth in 2016 Set Heat Record" [11] are some of the headlines appearing in recent years. Of the 17 hottest years on record, 16 have occurred since 2000. The 17th was in 1998.

Even more dramatic than the average temperatures are the extreme temperatures that have occurred. In 2006, temperatures reached 119°F in parts of Los Angeles and 121°F in Palm Springs. One hospitalized man had a body temperature of 109.9°. The heat wave was blamed for over 100 deaths [12]; more than 16,000 dairy cows also died from the heat. In Europe in 2003, a heat wave killed an estimated 35,000 people, and in 2010, an estimated 100,000 people died from a heat wave in Russia [10]. In India in 2016, temperatures reached 123.8°F and heat waves killed 2400 people. In Pakistan, 1200 people died from heat stroke [13]. In the Middle East, temperatures reached 129°F in Iraq and 129.2°F in Kuwait [14]. Human beings can't survive long at such temperatures, and neither can most animals, other than a few desert reptiles. "Stepping outside is like walking into a fire. It's like everything on your body - your skin, your eyes, your nose - starts to burn" [15].

In terms of comfort and danger, weather forecasters specify the heat index rather than the temperature alone. The heat index combines the raw temperature and the humidity. As the temperature increases, the body has to work harder to maintain its balance at 98.6°F. It does this by evaporating moisture. As the humidity rises, this becomes increasingly difficult and at some point the body can't keep up, resulting in a cascade of illnesses including heat cramps, heat exhaustion, heatstroke, hyperthermia, dehydration, and even cardiovascular and respiratory illnesses. Future climate warming could lead to thousands to tens of thousands of additional deaths each year from heat in the summer [16]. A *dry* temperature of 95°F is uncomfortable but livable as the body sweats to maintain itself, but with a high humidity, the heat index (what it feels like outdoors) can exceed 160°F [17] which nothing can survive for long. The Middle East is predicted to experience higher and higher heat indices over the 21st century [15, 17], causing many thousands of deaths and creating potential major refugee crises. Iran has already experienced a heat index of 163°F [13].

Another concept which has been developed by the National Weather Service is "Danger Days," when the heat index is 105°F or higher [18]. Previous years indicated that cities in the U.S. averaged only 1 danger day a year. Expectations are that by 2030, 85 cities and a third of the U.S. population will undergo at least 20 danger days annually, and by 2050, some areas in the U.S. will experience more than 100 danger days a year. The expectation, unless people stay inside nearly all day (impossible for many occupations like agriculture) is that many cases of heat exhaustion and fatal heat strokes will take place [18]. The number of days with a heat index above 90°F, known as Extreme Caution days, will also be much greater.

The rising global temperature does not occur uniformly, and some places will undergo global warming more than others, with significant resulting consequences. During 2016, parts of the Arctic were over 30°F higher than normal [11, 19], accelerating the melting of sea ice and adding to the positive feedback effect as the darker water absorbs more sunlight. "Temperatures are heading toward levels that many experts believe will pose a profound threat to both the natural world and to human civilization" [11]. If these excess temperatures are present in Greenland, the massive ice sheet could melt even faster and catastrophic sea level rise could occur. Temperatures are also above the global average in the Antarctic, accelerating ice sheet melt there.

"The pledges that countries are making to battle climate change would still allow the world to heat up by more than 6°F,.. a level that scientists say is likely to produce catastrophes ranging from food shortages to wide-spread extinctions of plant and animal life" [20]. If we were to actually use all the fossil fuels believed to be in the ground, some estimates are that the planet could warm by a staggering 16°F [21], leading to probable annihilation of most life, not the least because of all the refugee migrations and wars over resources that would take place before temperatures reached that extreme. Parts of the Earth may become uninhabitable within the 21st century, so hot in places, with or without humidity, that the body becomes unable to cool itself and heat stroke and death can occur in a few hours [17].

Animals suffer and die from the same extreme weather events as humans, particularly heat waves. Dogs and cats that have homes may be indoors along with their caretakers taking advantage of air conditioning, but stray animals and wildlife have no such option, and neither do the vast majority of farm animals. While some wildlife may be able to adapt to occasional high heat indices, especially if they have access to ponds, streams, or lakes to cool off, many will succumb to heat exhaustion and heat stroke just as humans would in the same circumstances.

Many species extinctions are expected to occur. Such widespread death of animals will only accelerate as conditions for their survival become ever more difficult. Wildlife deaths are seldom even mentioned; they occur in the forests where no one is present to witness them. Farm animal deaths are far more visible. In 1995, 14,000 cows and 1.8 million laying hens died in Iowa [22]. In the summer of 2003 when thousands of people died in Europe, many tens of thousands of farm animals also died. In the 2006 heat wave in the U.S. and Canada, 225 people died along with 25,000 cattle and 700,000 poultry, while in 2006 and 2007, heat waves killed over 2000 humans and a large but unknown number of animals.

Higher, dryer conditions also increase the populations of mosquitoes carrying West Nile virus, as well as anthrax spores, deadly to both animals and humans, and bark beetle infestations that destroy forests [23]. Animals suffer the same effects of heat and cold, emerging diseases, harsh weather, and heat waves as humans do but have far less ability to avoid these events [24].

Greenhouse Gases

The naturally-occurring greenhouse effect, and greenhouse gases (GGs) themselves, are not bad things. We owe our continued existence to the stable greenhouse effect that nature gave us; without this effect, there would be no life on the planet. Balancing the incoming solar radiation with outgoing infrared radiation and solar reflection in the absence of greenhouse gases, as if the planet had no atmosphere, would result in an Earth with $0^{\circ}F$ average global temperature. The oceans and all other bodies of water as we know them now would be frozen, and no plants could grow. Thanks to GGs, the average temperature has been a comfortable $59^{\circ}F$ for much of recorded history, with fluctuations of a few degrees plus or minus resulting from mini ice ages or mini warming ages. Starting with the industrial revolution, however, and the enormous amounts of fossil fuels burned, this balance has been tipped and the average temperature is creeping upward as already described.

The biggest contributor to global warming and by far most important GG is water vapor, the most ubiquitous greenhouse gas, contributing close to half of the total greenhouse effect. Water vapor transmits solar radiation at visible wavelengths but absorbs and re-emits infrared radiation coming up from the Earth, preventing its loss to space. Clouds also play a part in the greenhouse effect, absorbing or reflecting infrared radiation emitted by the Earth, and they also reflect some amount of visible sunlight back into space. On average, clouds contribute another 10% of the greenhouse warming effect.

The concentration of water vapor in the air varies daily, from humid to relatively dry conditions, and the humidity coupled with the temperature gives us the heat index already described. Carbon dioxide is also always present in the atmosphere, at concentrations established by a balance between sources and sinks. Forest fires, volcanos, human and animal expiration, vegetable matter decay, and burning fossil fuels represent sources, while trees, plants, and oceans represent sinks. Going back hundreds of thousands of years, this balance has resulted in an atmospheric concentration of CO_2 fluctuating around 250 - 280 parts per million (ppm). In the last 150 years or so, the burning of fossil fuels has raised this concentration to as much as 400 ppm. In the positive feedback effect already described, the increased CO_2 which has raised the surface temperature also increased the water vapor content since higher temperatures mean more evaporation.

Methane (CH_4) is another greenhouse gas emitted by decaying matter and by agriculture, especially animal agriculture, and by fossil fuel production, vegetation decay, landfill decomposition, rice farming, and permafrost melting. Methane's global warming potential is 25 to 72 times more effective than CO_2, and methane is eventually oxidized in the air to form CO_2 and water vapor, enhancing its greenhouse effect even more.

Nitrous oxide (N_2O) is yet another industrial greenhouse gas, coming from agricultural sources such as fertilizer use and crop burning. These molecules are several hundred times more effective than CO_2 in absorbing and re-emitting infrared radiation but are present in much smaller quantities in the atmosphere. Other greenhouse gases include ozone and man-made industrial gases such as the family of chlorofluorocarbons and hydroflurocarbons used for air conditioning. Most of these fluorocarbons are being phased out due to their damage to the protective ozone layer in the upper atmosphere.

The amount of CO_2 in the atmosphere has increased by over 38 billion tons a year [25], a number which is increasing. Other GGs are also on the rise. Burning fossil fuels is *the* major source, but deforestation is another major driver [26], contributing as much as 30% of global greenhouse gas emissions each year, especially if the trees are burned. Many forests have been and continue to be cut down to produce ranch land for raising cattle, who do their part in global warming by producing methane during digestion and breathing out CO_2. Forests are also cut down to clear land for the production of biofuels and palm oil. Gas leaks from the production of oil, coal, and natural gas also add substantial amounts of methane.

The oceans absorb a fraction of the CO_2 produced by these sources, and the absorbed CO_2 causes increased ocean acidification that damages coral reefs and destroys other marine life. *Re*forestation would recreate a major carbon sink with substantial benefits since forests can absorb an enormous amount of CO_2 each year [26], but at present the Earth is constantly *losing* forests, not *gaining* them. Added to mankind's deliberate forest destruction are the wildfires which are increasing in number and severity due at least in part to climate change and the increasing population of bark beetles which destroy large numbers of trees. Such infestations are also an increasing problem associated with climate change, mostly due to warmer winters which minimize insect deaths.

Politics are the biggest impediment to tackling the greenhouse gas problem, because politics, rather than innovation and technology, are largely determining our energy decisions. Burning fossil fuels has become the main source of electricity generation and the main source for transportation. The meetings of hundreds of nations in Rio, Kyoto, and Paris were mainly about curbing carbon dioxide emissions, but they had no real binding power. It will be necessary for the major industrial powers to agree on the severity of the problem and agree to take major steps together to limit the GW-CC effect, even though it may be too little and too late to save some of the developing island nations and coastal communities. There are many ideas, from cap and trade, taxing carbon emissions, increasing motor vehicle efficiency (gas mileage), or investing in renewable energy sources, to name a few. With an ever increasing population and bigger demand for energy per capita, the need for innovative solutions to the greenhouse gas problem is becoming increasingly urgent.

Wildfires

Every year the forest service and local fire departments battle wildfires that burn millions of acres of forest, destroying countless wildlife, destroying thousands of homes, and killing some of the incredibly brave firefighters who risk their lives to bring these fires under control. In 2015, 13 firefighters lost their lives, in 2014 the toll was 10 and in 2013 it was 34, nineteen fatalities in one location alone [27]. In 2015, 10 million acres of forest were burned [28] and the yearly number of wildfires has been increasing over the last 30 years. Some wildfires are accidentally set by careless hunters or campers while most are the result of lightning strikes or other natural causes. In any case, the increasing number of forest fires has been attributed at least in part to climate change [28-29]. The higher temperatures and drought conditions associated with climate change make the forest areas much drier and the forests more flammable. The number of wildfires each year, the severity and destructiveness of the fires, and the length of the "fire season" are all increased by climate change [28-31], and estimates are that climate change increases the number of wildfires that would otherwise take place by at least 50% [29-30].

These wildfires have a doubly-damaging effect. Growing trees represent a carbon sink. Destroying forests for purposes such as agriculture removes their ability to absorb CO_2 from the atmosphere, while burning forests put a great deal of CO_2 *back* into the air, carbon that was previously captured [32].

Just one wildfire in San Bernardino in the fall of 2016 caused the evacuation of 82,000 people and destroyed at least 100 homes. Wildfires routinely break out in Texas, New Mexico, Arizona, Nevada, Alaska, California, Colorado, and Utah. Dry conditions, hot temperatures, and strong winds all contribute, and are all worsened by climate change. In 2015, there were over 68,000 wildfires which burned 10,125,100 acres of forest. In 2014 there were 63,000 wildfires [33]. Wildfires cost as much as several billion dollars a year to fight in addition to the value of the lost homes and other property [34].

From a health standpoint, the smoke from these numerous wildfires can travel with the wind for hundreds of miles, smoke that represents fine particles that can enter and damage lungs and contribute to respiratory diseases [35]. Wildfire smoke contains particulate matter as well as carbon monoxide and dioxide, and significantly reduces air quality over large areas, increasing cardiovascular hospitalizations, emergency room visits, and increased asthma and bronchitis. It has been linked to hundreds of thousands of deaths each year across the globe.

Wildlife also suffers greatly from these wildfires. Besides breathing the same harmful smoke mentioned above, animals unable to move fast and far enough are destroyed by these fires. Another factor is habitat loss; the forests are their homes and supply

them food, water, and shelter, all taken away by these forest con-
flagrations. Wildfires, increased in number and intensity by climate
change, kills brave firefighters, causes the secondary health effects
and long-term fatalities of inhalation, costs many millions of dollars
in property damage, removes the forest benefits in acting as carbon
sinks, and kills countless animals of all types.

Sea Level Rise

The rising level of the world's oceans is deceptive, especially for
those people who don't live near the coasts. It may only amount to a
few centimeters, an inch or so, *per year*, hardly noticeable by itself.
However, this increase adds up over the decades, and when sea water is pushed on shore by storm surges or other extreme weather, the
water can intrude across sea walls, onto roads, lawns, houses, city
streets, subways, and coastal lakes and ponds, turning fresh water
into brackish and potentially damaging the fresh water supply that
countless people depend on. Sea level rise could become a catas-
trophe of world-changing proportions, destroying whole island nations, and making some coastal cities uninhabitable.

The rising global temperature contributes partially to this rise
due to thermal expansion [36]; water, like most substances, expands
as it becomes warmer. A potentially much larger contribution comes
from the vast amounts of water tied up as ice in the far northern and
southern regions, particularly the ice which sits on top of land. In the
Arctic, much of the ice is floating on the sea and partially melts in
the summer and refreezes in the winter. For the last several decades,
the net amount of sea ice has been declining at more than 10% every
decade [37-38], a testament to the increasing temperatures in the far
north, rising by an average of 11°F while the rest of the planet has
risen by an average of 2°F [39]. Fortunately, sea ice is already floating on the sea and therefore contributes little directly to the sea level

rise when it melts.

The far greater potential contributor to sea level rise is the melt-ing of land ice such as glaciers and the vast land ice sheets that cover parts of the world such as Greenland and Antarctica. As these glaciers and ice sheets melt, the melt-water runs down to the sea, adding directly to sea level rise. In other cases, huge portions of ice sheets and glaciers at the edge of the land sometimes break off into the sea, increasing the numbers of iceberg hazards, and also adding to sea level rise. Glaciers building in size for millennia have begun to shrink and "retreat" away from the coast as they lose ice, while others can accelerate their march to the coast as they become "lubri-cated" by melt water which finds its way between the bottom of the ice sheet and the land.

Hundreds of glaciers in Alaska have been retreating [40-42]. In Austria and the Alpine mountains, over 900 glaciers have been melting, raising the possibility of floods, mudslides, and tsunami-like devastation [43-44]. In the Andes regions of South America, glaciers melting at accelerated rates means loss of drinking water, loss of water for crop irrigation, and less hydroelectric power [45]. In China, 46,000 glaciers in the Himalayas and surrounding moun-tains are shrinking, endangering the water supply for 10 countries in Asia [46]. Most of the planet's glaciers are on the decline, and if global warming were to continue unabated, many if not most gla-ciers would cease to exist, causing major problems for millions of people [46-47].

Glaciers melting all over the world could contribute several feet of sea level rise, but this pales in comparison to the effect of the massive ice sheets of Antarctica and Greenland. Already the projec-tions for sea level rise by the end of this century have been doubled because of the melting Antarctic ice sheet [48]. The Greenland Ice Sheet is melting even faster, losing up to 300 billion tons of ice a year [49], and if it were all to melt, the sea level could rise 23 feet

[50]. However, the Antarctic ice sheet contains 10 times as much ice as Greenland, and if it were to all melt, it would raise sea levels by a staggering 200 feet [51]. The U.S. Geological Survey [52] puts the number even higher, 240 feet if all of Antarctica were to melt, and 265 feet if all the ice on the planet were to melt. Also, as the ice melts, it adds fresh water into the ocean, potentially interrupting the warm ocean current "conveyer belt" that moderates the climate of the Northern Hemisphere [53], as described earlier.

What are the possibilities that such melting could happen? Studies have shown that if all the fossil fuel reserves now in the ground, coal, oil, natural gas, were to be extracted and burned at present rates, the CO_2 concentration in the atmosphere would rise a factor of 4 to 8 and all the ice on the planet would be gone [54-55]. Temperatures would rise at least 16°F and sea levels would exceed 200 feet beyond today. Civilization would be greatly endangered as humans made war over diminishing water, food, and habitable land. The sweltering temperatures alone over most of the year would kill most life forms. Even a sea level rise by 5 or 6 feet, a possibility for the end of the 21st century, could inundate countless coastal and some inland communities, including Florida, Louisiana, New York, Miami, London, Venice, Hong Kong, Sidney, Houston, Washington DC, Amsterdam, Paris, Berlin, Rome, Tokyo, Bangladesh, and many of the Pacific Islands [56-61].

These frozen ice sheets are home to gigantic reservoirs of ancient microbes frozen in the ice for hundreds of thousands of years [62], microbes that humans and animals have had no exposure to and no immunity from. As the ice melts, some of these microbes will be exposed to the environment and could represent new types of diseases. Microbes frozen in the ice for 750,000 years have been brought back to life, and scientists believe that the same could be true for microbes frozen for millions of years [62]. Melting ice sheets and glaciers also expose volumes of decomposing organic matter that contributes methane and carbon dioxide to the atmosphere, another positive feedback effect that exacerbates global warming.

Human beings in coastal communities and low-lying islands may be able to migrate as the sea rises, but the wildlife in many communities doesn't have such an option, particularly on islands. As the sea rises, available habitat shrinks, forcing wildlife into ever-smaller spaces and increasing competition for the remaining food and fresh water. If the islands become totally inundated, as predicted for some, the animals on those islands will perish.

Climate change meetings such as Kyoto, The Hague, Rio de Janeiro, and Paris, often contain pleas from island nations to curtail greenhouse emissions emanating from major industrial countries, pleas which fall on deaf ears. China, India, the U.S., and others apparently intend to continue burning fossil fuels as usual with the consequent rise of CO_2 concentrations, while other nations lose their homes, refugee crises are created, water shortages turn into crises, and countless animals die that are unable to adapt. Already, the iconic polar bear is suffering from retreating sea ice, as his prey food disappears along with the ice. Polar bears are in danger of significant population decline, if not extinction [49, 63-65].

Extreme Weather.

Heat waves, floods, wildfires, gales and hurricanes, thunderstorms, downpours, and droughts; these are some of the events coming under the category of extreme weather. People might not feel the rise in average global temperature of a degree or two, but they will surely feel the effects of more violent hurricanes, intense blizzards, destructive floods, and debilitating heat waves. It has already been described how heat waves are increasing and costing many thousands of lives, and how wildfires burn millions of acres each year and cost billions of dollars in damages as well as costing human lives, including the second-hand effects of smoke pollution down-stream resulting in long-term health effects. Countless animals also suffer and die from these heat waves and forest fires.

Scientists are the first to say that no single weather event can be attributed to a single cause; hurricanes, thunderstorms, heat and cold waves, etc., would still occur without man's presence and influence. However, *trends* in the increase of extreme weather over periods of time have been attributed to human activities with a high degree of probability [66]. The number, duration, and intensity of these phenomena can be attributed to man's influence on the climate, to the widespread burning of fossil fuels and consequent dumping of billions of tons of greenhouse gases into the atmosphere.

Heat waves as described earlier are the leading weather-related cause of death in the United States, due to heat stroke, cardiovascular and cerebrovascular disease, and respiratory diseases [66]. Areas of the Earth may become uninhabitable, leading to increasing refugee crises and food, water, and basic necessity shortages. It's virtually certain that climate change is responsible for extreme heat waves [66, 68], and in many places on the globe, hotter, drier conditions with many days above 95 - 100°F may be the new normal, spreading famine, sparking food and water riots, and destabilizing governments [67].

Hotter, drier conditions also dry out the soil, impacting agriculture and necessitating more irrigation which increases the burden on water resources, at a time when water supplies may be compromised by reduced rainfall and reduced melting snow pack. Some locations will see strongly increased rainfall and flooding while droughts become worse in others. Crop yields are expected to decrease [66-67, 69], even as the expected human population will expand in the next century, greatly increasing food demand. Deserts are expected to grow in area [70] as droughts become worse, encroaching on important agricultural land. In China, the Gobi Desert has been expanding at a rate of 1350 square miles a year, shutting down several villages and causing the relocation of thousands of people [71]. In Africa, one of the world's largest artificial lakes has almost run dry, with the consequent failure of its associated Kariba Dam that has been supplying hydroelectric power to several countries, causing blackouts and setting back economic development [72]. The drought in the western United States that began in 2011 was made worse by global warming while at the same time the snow pack providing valuable water was at record low levels [73]. Short periods of high rainfall are followed by longer periods of drought in parts of the U.S., both expected to increase due to climate change [74]. The Amazon rainforest had its second 100-year drought in 5 years recently, and 20 countries set new records for extreme heat [67].

Floods could be called the opposite of droughts. Around the world, from 1980 to 2009, floods caused more than 500,000 deaths [66], and floods are expected to increase in frequency and severity due to global warming [66, 70, 75]. In Pakistan, floods displaced 20 million people [67]. Both sea level rise and extreme rainfall events contribute to flooding. In Louisiana in 2016, torrential rains led to deadly floods that killed 13 people, destroyed 55,000 homes, and damaged 6000 businesses with 9 billion dollars in damages [75]. The phenomenon known as an "atmospheric river" can deliver much needed water to parched areas but cause megafloods in others. In an atmospheric river, large quantities of water vapor flow through the atmosphere a mile above the ground and can cause torrential rain when they encounter land. A flood in 1861 deluged California with 10-15 Mississippi River volumes of rainfall, causing an inland sea 300 miles long and killing thousands of people and an estimated 200,000 cattle [76]. Megafloods like this occur naturally every 200 years, but smaller floods caused by this phenomenon happen frequently and are expected to increase in number and intensity [76]. Floods predicted by climate models could lead to the evacuation of

1.5 million people and cost $400 - $700 billion in damages [76].

Hurricanes are also expected to increase in intensity and duration due to global warming [66, 77–80]. Greater moisture content in the atmosphere due to higher air temperatures and warmer sea water are the driving forces for hurricane formation, and the higher intensity and duration will increase coastal flooding events and flooding due to storm surges. Gale-force winds and torrential downpours, slightly less virulent than hurricane strength, are also predicted to increase in number and intensity.

Climate change deniers often claim that efforts to minimize global warming's extreme weather events will be too expensive. What's usually left out of the debate is the very high costs of damage from these events if efforts aren't taken to prevent them. In addition to the many thousands of human fatalities, the extinction of species, the increase in many human diseases, and the damage to water and food resources, the costs of mitigating and adapting to climate change once it takes place are likely to cost many billions of dollars [67, 70, 75-76, 80-81], and money spent to minimize such events is a good return on investment. "Global losses from extreme weather have quadrupled from the 1980s to $200 billion a year. ... Recent years have brought raging fires that have devastated the West, prolonged droughts that have wilted the Southwest, and fierce storms that have inundated the East Coast" [81]. The Louisiana flood damaged 60,000 properties and cost an estimated $30 billion [80]. Hurricane Sandy in 2012 cost an estimated $65 billion in damages [78]. The costs of repairing devastation from climate change will very likely significantly exceed the costs that today's politicians claim would be needed to prevent it.

Some of the many costs that will come from not preventing or minimizing climate change include damage to coastal cities, damage to agriculture, costs of treating increased disease outbreaks, lost productivity from those who succumb to extreme heat waves, losses of homes, businesses, and natural habitat from increases in wildfires, losses from floods and droughts, and damage from major storm events like Hurricane Katrina and Hurricane Sandy. The damages and costs of these climate change consequences are seldom mentioned in political circles. Neither is the enormous toll on wildlife caused by extreme weather, or even the many human fatalities that can take place.

Agriculture

A major concern for humans and animals alike is what climate change and global warming will do to the food supply. The present human population of 7 billion people is expected to rise to more than 9 billion by 2050, with a consequent demand for food and water resources [82]. A large fraction of the Earth's available land area has already been used for agriculture, 30% of the ice-free land surface [83], 70% of which is used for livestock production. Although there is more forest land still left that some might argue could be put into use, this would be disastrous for the planet, as forests sequester enormous amounts of carbon both in the trees and the soil, all of which would go into the atmosphere as CO_2 whether burned or not. (The Amazon and other rainforests are particularly large carbon sinks because of their lushness. 70% of previously forested land in the Amazon has already been cut down for pasture land to raise cattle [83-84].) We need to plant *more* forests, not reduce them; we are already losing forests to desertification from increasing droughts as well as clearing for pasture land.

Since we don't have more land to make use of without paying a strong penalty, greater crop yields would be a blessing: more productivity from the land already in use. Higher CO_2 can actually increase yields of some crops [85-88], acting as a kind of fertilizer aiding in photosynthesis and requiring slightly less water. Plants are known to grow faster in high CO_2 environments. Unfortunately, the higher CO_2 can also lower the nutritional value of many crops. The carbohydrate content of plants may increase, but the protein and mineral content declines, reducing the plant "quality" as a food source [87, 89]; in particular, the nutritional value of important crops such as wheat and rice will decrease as protein and essential minerals are reduced [89].

Another factor affecting food quality is atmospheric temperature. All crops have a narrow temperature range for optimum growth and optimum seed production, and have temperatures above which the crops will neither grow nor reproduce. For example, corn fails to produce seeds at temperatures above 95°F and soybeans at temperatures above 102°F [88], both of which are expected in many parts of the world as the Earth heats up. While *average* temperatures have risen between 1 and 2 degrees Fahrenheit, *local* temperatures can fluctuate 5-10°F as heat waves occur. Crop yields are expected to fall as average temperatures rise [69, 82, 88, 90], offsetting the effects of increased CO_2. In some cases, planting will need to move northward to higher latitudes as temperatures become higher and climates dryer in southern zones, [69, 86].

Estimates range from a loss in corn and wheat yield by 5-15% for each degree of temperature rise in the U.S., Africa, and India [90] to as much as a loss of 40% for crop yields globally in coming decades [67]. Some fruits such as apples and berries require extended low temperatures in winter to produce abundant yields the following summer, but rising temperatures make this increasingly unlikely [88]. Droughts, floods, and heat waves will have a detrimental effect on crop yields [86, 88], decreasing the food supply even as the demand increases.

Warmer climates, higher humidity, and higher CO_2 all increase the amounts of weeds, insects, bacteria, parasites, and fungi that detrimentally affect plants and livestock. Farmers will then use higher amounts of chemicals - pesticides and herbicides - on crops fed directly to people or to farm animals, increasing the human exposure to chemical pollutants in their diet [83, 87-89]. The increased use of chemicals represents a growing threat to biodiversity, while climate change in all its aspects: heat waves, droughts, storms, pests, alien species, and chemical use, is a major contributor to species extinction [83, 91-92].

Just as climate change affects agriculture, agriculture in turn contributes to climate change. Early estimates suggested that at least 18% of global greenhouse gas emissions were contributed by agriculture, more than the entire transportation sector (cars, trucks, railroads, airplanes) [83], but this already high number omitted the large contribution coming from cutting down forests in order to create pasture land for cattle grazing and raising feedcrops for livestock. Deforestation adds as much as 80 tons of carbon per acre into the atmosphere when forests are first cut and another 80 tons from the soil per acre as times goes by [93]. Adding these factors suggested that 30-35% of global greenhouse gases are contributed by agriculture [82, 94] and after adding respiration from livestock (which exhale CO_2), methane from manure and livestock flatulence, fertilizer production, fish aquaculture, and all the energy needed to carry out these activities, as much as 50% of all global greenhouse gases has been suggested to come from food animal production [93, 95], when all the factors involved with animal agriculture are considered.

In addition, raising livestock is a major contributor to serious environmental problems, including the already mentioned deforestation, desertification in dry areas suffering from drought, animal waste storage and disposal, water pollution with pesticides, antibiotics, and heavy metals, and loss of natural ecosystems (biodiversity) due to habitat loss [83, 94, 96]. This habitat loss is a major cause of endangered species and species extinction, said to be 50 to 500 times the *natural* rate of species extinction today [83].

The good news in all this is that animal agriculture represents one of the most potent methods to reduce the climate change problem. Reducing worldwide consumption of animal products would be a powerful way to reduce global warming and climate change and would result in many health benefits, reducing cardiovascular disease and colorectal cancer [82-83, 94, 97-98]. It would also reduce both the amount of land and the amount of precious fresh water needed for agriculture and could contribute to reversing deforestation [82, 83, 93, 96].

Ethics

"Climate change will cause harm. Heat waves, storms, and floods will kill many people and harm many others. Tropical diseases, which will increase their range as the climate warms, will exact their toll in human lives. Changing patterns of rainfall will lead to local shortages of food and safe drinking water. Large-scale human migrations in response to rising sea levels and other climate-induced stresses will impoverish many people. As yet, few experts have predicted specific numbers, but some statistics suggest the scale of harm that climate change will cause. The European heat wave of 2003 is estimated to have killed 35,000 people. In 1998 floods in China adversely affected 240 million. The World Health Organization estimates that as long ago as [the year] 2000 the annual death toll from climate change had already reached more than 150,000." So states John Broome in an article entitled "The Ethics of Climate Change," (Scientific American, June 2008). He asks the question of how people alive today should consider the lives and well-being of those living a few hundred years from now that will have to deal with the devastation that the present generations are leaving behind.

What could be added to this article on ethics is the probability of mass conflicts between people fighting for dwindling resources - food, water, living space - in a world with increasingly uninhabitable land area and increasingly unlivable local temperatures. It also leaves out the devastation to the *non-human living beings* on the planet, who suffer from those same floods, storms, wildfires, and debilitating temperatures as people do. And if humans probably aren't in danger of becoming extinct, that isn't true for many species in the animal kingdom.

Millions, probably hundreds of millions, of people will become refugees, migrants, as the local habitat becomes unable to sustain them or rising sea levels inundate their living space. Small numbers of refugee people may be welcomed by others in the world community, but what will happen when entire populations need to migrate onto land occupied by others, diluting and placing severe burdens on the present inhabitants [99]?

If all this sounds like science fiction and implausible, think again; it's the simple consequence of a runaway greenhouse effect if we humans don't do something to change the path we're on. It may be a very pessimistic prediction, but it's suggested by solid and unassailable science and confirmed by many of the climate/weather events that have taken place in the last 30-40 years.

The good news is it doesn't have to be this way. Prevent climate change as much as humanly possible and all these negative consequences go away. There will still be storms and heat waves, droughts and hurricanes, but only the much smaller number that occur naturally, and those we are already used to.

HELPING PEOPLE AND ANIMALS.

Many of the effects of climate change on harming people's health and well-being have already been described. Wildlife and domestic animals will also suffer from these effects, more than people will. Wildlife has no hospitals, air conditioning, or antibiotics to call upon.

Wildfires have been increasing in frequency and intensity, as described earlier. Global warming is a major contributing factor to wildfires [100]. Fires are deadly to wildlife. Some like birds may be able to fly out of danger but they leave nests, eggs, and young behind. Large mammals like elk and deer may be able to escape but many small animal species are unable and will perish [101-102]. Young animals and small animals are most at risk from fires [101, 103] and even fish and amphibians are at risk as ponds and streams become too warm and polluted with debris [103]. Invertebrates that can only crawl out of harm's way will almost surely perish [102].

Those animals that survive may not be so lucky either. When their habitat is destroyed by fire, finding food, water, and shelter becomes a difficult and stressful task [104-105]. Unburned habitat may be shrinking and less available due to the sheer number and magnitude of wildfires that destroy millions of acres of habitat each year [103], creating great hardship for animals trying to survive. New habitat is also likely to be already occupied, creating competition for available resources and possibly exceeding the carrying capacity, leading to conflict and possible starvation.

Heat waves are just one of the natural disasters that both humans and animals face, but while people can stay indoors, use air conditioning, and have ample food and water, animals don't have these luxuries. In many countries of the world, luxuries like air conditioning are too expensive and unavailable even to many millions of people. According to the NRDC (Natural Resources Defense Council), "in the United States, hundreds of heat-related deaths occur each year due to direct and indirect effects of heat-exacerbated, life-threatening illnesses, such as heat exhaustion, heatstroke, and cardiovascular and kidney diseases. Indeed, extreme heat kills more

Americans each year, on average, than hurricanes, tornadoes, floods, and lightning combined" [106].

Many more animals die from extreme weather events like heat waves than humans do, from the inability to escape the severe conditions [107]. Millions of animals of many species die unseen and unheard in the forests, deserts, and mountains. Animals with fur can't shed their "clothes" to feel cooler; many can't sweat as humans do in the human body's natural cooling mechanism. Birds can only cool off by panting with their beaks wide open, and many die from dehydration as they lose moisture this way [108-109]. Many mass die-offs of birds have occurred. Habitats may become uninhabitable as temperatures rise, leading those species and individuals that can to migrate northward or to higher elevations. This potential future isn't just true for animals; parts of the planet may become too uninhabitable for people as well [110], causing the conflicts, refugees, and migrations that have been alluded to before.

Australia has been in the forefront in the study of the effects of climate change on wildlife, since Australians monitor what's going on in the environment much more consistently than most other countries. Global warming and climate change in Australia have been monitored for years, and the country suffers from many heat waves [111-113] which are very damaging to its wildlife, from giant fruit bats known as flying foxes to kangaroos to possums and koalas. Massive die-offs of wildlife have occurred, including 100,000 flying foxes over a period of just a few days. "It is clear that we are witnessing a great (perhaps the first great) mammalian mass death event that is directly attributable to global warming. This is a vision of the world that is coming - the flying-foxes are among the first to fall, but they will not be the last" [113]. Mass wildlife die-offs have taken place over much of Southeast Asia and in the U.S. as well [114-115].

Livestock all over the world are also at significant risk from excess temperatures, with the young, sick, pregnant, or overweight animals the most vulnerable [116]. Most species have a maximum tolerable temperature, and when the ambient temperature approaches or exceeds that value, hyperthermia sets in, organs begin to fail, heat stroke can occur, and dehydration can take place. Humans have the same physiological response, but can escape the heat to more comfortable conditions, at least in some countries. When they can't, mass human casualties take place.

In 1995 in the central U.S., an unknown but significant number of people died along with thousands of cattle and 1.8 million laying hens [117]. In July 1999, 3000 cattle died in Nebraska and in the year 2000, in Australia, 24 people, 2000 cattle, and an undisclosed number of dogs, horses, and wildlife died. When the 35,000 people died from the heat wave in Europe in 2003, hundreds of thousands of animals of many species also died. In the California heat wave in 2006, 225 people died along with 25,000 cattle and 700,000 chickens and turkeys [118]. In Europe that same year, more than 2000 people died and untold numbers of animals. In parts of Asia, heat waves and human and animal deaths have become routine; India experiences thousands of people and animals dying nearly every year [119-120].

Droughts cause life-threatening problems for wildlife. The high temperatures dry out the land (leading to more wildfires as already discussed), cause poor growth of vegetation and crops, and cause water shortages [121]. Many wildlife mass deaths are related to these food and water shortages, which can also lead to the deaths of many farm animals [107]. In some cases, entire lakes have dried up, leading to major water crises [114]. Some animals are able to migrate during droughts and heat waves, finding more suitable and hospitable habitat. Birds are particularly able to acclimate by flying to better areas, but many species can't adapt as easily and are more likely to perish [122-123].

*Flood*s can result from storm surges or torrential downpours or may come from high volumes of water accompanying hurricanes. Human deaths may come from physical injury such as building collapse or from drowning, while wildlife and livestock deaths are almost always associated with drowning. Wildlife is also destroyed in the aftermath by loss of livable habitat as flood waters and storms deteriorate many square miles of land. Extreme rainfall is becoming more frequent and intense due to climate change, causing rivers to overflow, drowning countless animals directly or destroying their food sources and their habitat. Ground-nesting birds and animals, invertebrates, the young and newly born, and others perish from overflowing and often polluted rivers, streams, ponds, and lakes and from the runoff from surging and receding waters.

Thousands of people have lost their lives in storms, hurricanes, and the associated floods. Hurricane Mitch in 1998 killed at least 9,000 people with another 9,000 missing [124]. Hurricane Jeanne in 2004 killed 3000, and Katrina in 2005 killed at least 1800 [125]. Hurricane Matthew in 2016 killed 46 in the U.S., and killed at least 1000 people in the Caribbean and sparked fears of a cholera epidemic from contaminated water [126-128]. The cost in damages from these storms totals many hundreds of millions of dollars. While these thousands of people died from these killer storms, literally millions of animals were killed at the same time.

Reliable estimates of wildlife casualties from extreme weather are difficult to obtain since they are scattered over large areas, but the death toll of domesticated animals, primarily livestock, is relatively better known. Hurricane Floyd in 1999 drowned at least 10,000 pigs. Hurricane Matthew and its associated floods are estimated to have killed 5 million chickens and turkeys, along with countless pigs, in North Carolina alone [129]. The Fort Bend flood in 2016 killed "countless numbers" of cattle, horses, goats, sheep, chickens, and dogs [130].

Factory farming conditions with livestock confined in large warehouse-like buildings make it difficult for animals to escape when flood waters inundate these buildings. To make matters worse, flood waters infiltrate into the manure lagoons that accompany pig farms and overflow into rivers and streams, destroying any wildlife that depends on these waters and possibly contaminating water that humans depend on as well [129]. Family pets are at great risk also, left behind to drown or starve when people are rescued [107]. During hurricane Katrina, an estimated 600,000 companion animals drowned or starved when they were denied access to shelters along with their human families [107].

Storms and hurricanes would take place naturally without climate change, and no one storm can be attributed directly to global warming. Looking at the historical record however, the number and severity of extreme weather events has been increasing, and this *can* be attributed to climate change. The human and animal deaths from any one storm may not be blamed on climate change, but thousands of people and millions of animals have perished and will continue to perish from storms and floods and heat waves that wouldn't have taken place without climate change.

Extinction of the earth's precious wildlife has already been described in relation to habitat loss, hunting, and government wildlife policies, and most of all by the horrendous bushmeat trade which is wiping out whole populations of animals. As bad as such extinctions are, they are much smaller than the extinctions that are already being brought about by climate change. All the consequences of climate change already described - heat waves, wildfires, droughts, floods, increasing global temperature, extreme storms, higher sea levels, melting sea ice - make it difficult for many vulnerable species to survive. It is said that the Earth is in the midst of the 6th mass extinction [131], the first since the dinosaurs succumbed 65 million years ago [132-134]. Pollution, deforestation, bushmeat, but most of all climate change are driving birds, fish, amphibians, invertebrates, marine and land mammals, and reptiles onto the endangered species and "already extinct" lists. The planet is losing species at the rate of dozens per day, between 1,000 and 10,000 times the natural extinction rate [134-135].

Estimates of species loss over the next 50 to 100 years vary widely but all represent catastrophic extinction, ranging from 15% [136], to 1/6 [132, 137-138], to 25% [139-140], to 50% [134] of all known species. The differences in these estimates come from different projections on the rate of fossil fuel burning and consequent temperature rise and extreme weather conditions. Many species not driven extinct would still experience substantial detriments in habitat, food and water resources, disruptions to behavior patterns like mating and hibernation, and damage to sustainable functioning ecosystems that they need to survive [138, 141].

Some species will be able to migrate to cooler habitats, usually farther north or higher elevations, preventing their extinction [123, 142-143], but the rapid pace of climate change is likely to exceed the ability of many species to either migrate or adapt [139]. One example is the iconic polar bear, dependent on the disappearing sea ice for food and survival [144-145]. Many polar bears are starving due to their inability to adapt quickly enough [144] and the species could become extinct in 100 years or less [139]. Adelie penguins, amphibians, reptiles, many species of invertebrates, several species of sea turtles and whales, and some bird and fish species are also seriously endangered due to climate change [142,146].

Corals are another example of a species unable to migrate, expand their range, or adapt to climate change. All over the Earth coral reefs are dying due to human influence, from chemical pollution, coastal sediment and agricultural runoff, to rising water temperatures, and rising acidity [144, 147-149]. A great deal of marine life depends on healthy reef ecosystems to breed, feed, and survive

[147]. The Australian Great Barrier Reef, visible from space, is in serious danger, with mass bleaching along 1400 miles of its length and as much as two-thirds dead or dying [148, 150-151]. Australia depends on the reef for 70,000 jobs and billions of dollars in tourism [151], much of which is in increasing jeopardy.

Diseases: Human Health and Animal Health

Animals get sick just like people do, suffering from parasites, viruses, and bacteria just like humans do. Animals can suffer arthritis, diabetes, heart disease, strokes, and cancer. Their health and longevity depends on the strength of their immune system just as in the human body.

Sometimes the illnesses that plague animals are the same as those affecting humans; sometimes they are different. Dogs fall ill and die from heartworm for example, while people don't suffer from heartworm even if bitten by the same mosquitoes. But there are many diseases and pathogens that attack both animals and humans alike, and the list is quite long: rabies, ebola, polio, malaria, cholera, plague, Lyme, West Nile virus, influenza, encephalitis, SIV, tuberculosis, staphylococcus aureus, salmonella, Giardia parasites, brucellosis, ehrlichiosis, hantavirus, measles, Zika, Marburg, yellow fever, and small pox, among others. Not all species get all diseases. For example, there are 200 varieties of the malaria parasite; humans are subject to 5 of them while animals succumb to others [152], but one strain overlaps animals and humans. Birds, antelopes, bats, and lizards all suffer from malaria. A quarter of the U.S. deer population is said to be infected with malaria [153].

The great apes share much of the same DNA as humans and it's no surprise that they share the same diseases, including the deadly Ebola and Marburg viral infections [154-157]. Thousands of gorillas have been killed by ebola, in some cases nearly entire populations

[158]. This goes on at the same time as the human ebola epidemics are taking place. Gorillas, chimpanzees, and bonobos are also susceptible to well-known human diseases such as measles, pneumonia, influenza, staph infections, and the simian form of HIV (SIV) [157, 159-161]. The human form of immunodeficiency virus may have come originally from eating infected monkeys or apes in the bushmeat trade [162]. Other viruses in the HIV-SIV family can infect sheep, goats, horses, cattle, and cats [161].

West Nile virus can cause fatal neurological disease in some humans (though symptoms can be mild in others) but is also deadly to many bird species [163] and to many other species including horses, cats, dogs, chipmunks, skunks, squirrels, rabbits, and alligators [164]. It is responsible for 1000 human fatalities and millions of animal fatalities each year [165]. Droughts, made worse by climate change, increase the incidences of WNV [166]. Dengue fever, once rare in North America, is migrating northward into the U.S. from South America [165] and is estimated to infect 50 million people worldwide each year with 10,000-12,000 deaths [167], with some estimates going as high as 50,000 deaths [168]. The disease came originally from monkeys in Southeast Asia.

Cholera, a bacterial disease that caused deadly epidemics in Europe and still infects many people worldwide, can also infect bison, cattle, and dogs [169]. Outbreaks in Latin America, Africa, India, and the Middle East often follow heavy rainfall and floods [165, 170-171]. Heavy rains and floods, whether related to periodic El Ninos or not, contribute to outbreaks of malaria, dengue, encephalitis, hantavirus, and Rift Valley fever as well as cholera [170]. Bubonic plague, the disease that killed off a third of the population of Europe in the middle ages and half the European population in earlier years [172] is still found in many parts of the world including Africa, Asia, North and South America and the Middle East, and also infects 200 animal species including rodents, squirrels, bobcats, coyotes, goats, camels, sheep, dogs, and cats [173].

Tuberculosis, the lung disease that was a leading cause of death in the U.S. and Europe in the 1800's, still infects a third of the global population today and kills 1.5 million people a year [174]. The human strain, mycobacterium tuberculosis, and the animal strain, mycobacterium bovis, are very similar and humans can contract the disease from infected livestock and particularly from unpasteurized milk or cheese. Many famous people contracted TB and many died from it, including Edgar Allen Poe, George Orwell, Frederic Chopin, Andrew Jackson, Eleanor Roosevelt, Henry David Thoreau, John Keats, and Emily Bronte. Many species of animals are susceptible to TB and have died from it including horses, sheep, goats, cattle, pigs, deer, dogs, cats, elephants, and buffalo [175].

Given that it is established that humans and animals suffer from many of the same diseases or close strains, what has this to do with climate change? The answer is that many of these diseases are transmitted by vectors: mosquitoes, fleas, ticks, black flies, and populations of these are strongly influenced by climate conditions. The pathogens themselves that cause the diseases: the viruses, bacteria, and parasites, are also strongly affected by climate conditions. Climate-sensitive diseases are said to be among the planet's largest killers, causing millions of deaths each year [176]. (As discussed earlier, pathogens, hosts, and vectors would be still be present naturally without climate change, but climate change makes the populations and incidences of diseases that they spread much worse.) Floods spread bacteria and viruses as well as chemical pollutants and increase mosquito populations. Droughts, paradoxically, can also increase mosquito populations [166, 176] as well as populations of rodents that are the disease reservoirs [176].

Warmer atmospheric temperatures increase the populations of the hosts and vectors, as well as the pathogens, and shorten their incubation cycles, while warmer winters and earlier springs *increase the length of the time period* that the vectors and pathogens can spread disease [88, 165 176-178]. Warmer temperatures also increase the number of pathogen *generations* per year [177,179], and expand both the range and migration north and south of mosquitoes and ticks [176, 178-180]. Diseases such as malaria, Lyme, encephalitis, yellow fever, plague, and dengue have all increased their active area and expanded into higher latitudes, accompanied by their mosquito, tick, and midge vectors [88, 179, 181-182]. Chikungunya is another mosquito-borne viral disease that can cause excruciating joint pain and has been migrating northward into the U.S. as temperatures rise [165, 180, 183]. A parasite that damages and kills oysters has expanded more than a thousand miles along the U.S. East coast due to warmer temperatures [182, 184], degrading both the oyster population and the oyster industry.

Corals are endangered world-wide. In addition to the bleaching and deaths due to pollution, warmer temperatures, and acidification, warmer waters also increase coral diseases from viruses and bacteria, which have expanded in range across the globe in response to climate change [179, 185]. Some of these same viruses, detected in coastal waters, have been linked to human diseases including heart disease, diabetes, meningitis, and hepatitis [185].

At the same time that climate change is increasing pathogen population, range, and virulence, it also weakens the immune system that all of us, human and animals, depend on to fight off infection and disease [186-187], leading potentially to shorter lifespans, more severe infections, and sometimes cancer. Heat waves, droughts, storms, and rising temperatures all appear to affect this life-saving immune competence and increase the risk of disease [187]. Marine mammals, corals, and other marine organisms are also vulnerable to environmental effects on their immune systems [182,187], and amphibians (frogs, salamanders, toads, newts) are particularly susceptible [181].

Humans can acquire many types of diseases. Animals can acquire many types of diseases. Climate change, by its exacerbating effects on hosts, vectors, and pathogens, can make all of the diseases more severe, more frequent, and more wide-spread, over and above the naturally occurring rate that would exist without climate change. Many excess human and animal lives will be lost to these ever-worsening effects.

Diet and Climate Change

The effect that animal agriculture and human dietary preferences have on global warming and climate change has already been briefly described. It's already widely known that a diet that includes an excess of meat can have human health consequences. Obviously, farm animals pay a heavy price for humans' dietary preferences.

There are between 50 and 70 billion farm animals raised and slaughtered every year [95-96], many of them enduring the horrors of factory farming where "chickens, pigs, turkeys, and other animals are confined in cages, crates, pens, stalls, and warehouse-like grow-out facilities ... devoid of environmental stimuli, adequate space, or means by which to experience most natural behaviors" [96]. Whether it is newborn calves confined to tiny veal crates, multiple laying hens in cages without room to even turn around, pigs in gestation pens unable even to move, newborn male chicks thrown into garbage cans to slowly suffocate, or terrified horses being transported to slaughter, farm animals never have a nice day. That an enlightened society of humanitarian people would allow this to go on year after year seems hard to understand, and this doesn't even include the horror and terror of the slaughtering process. Obviously changing this treatment of farm animals would help them, and it would help people as well.

Many medical studies using epidemiology have been carried out to study the effects of diet on human health, taking into account the "confounding" factors such as age, race, smoking, alcohol, weight, family history, exercise, pre-existing conditions, and so forth. In one study of 500,000 people, the health and longevity of vegetarians and non-vegetarians were compared for red, processed, and white meat intake [180]. For overall mortality, red meat increased men's risk by 31% and women's by 36% while processed meat raised mortality risks by 16% for men and 25% for women. For cancer specifically, the risks were also higher by around 20% for red meat eaters and 12% for processed meat, in both men and women. For cardiovascular disease (CVD), the biggest effect was seen in women, with a 50% and 38% higher risk for red and processed meat, respectively. White meat had a small or even slightly beneficial effect in the study.

Other studies confirmed the benefits of low or no-meat diets. In a study which included participants from Germany, California, the Netherlands, Japan, and the UK, vegetarians had a 29% lower mortality risk for heart disease, an 18% lower risk for cancer, and a 12% lower risk of stroke [188-189]. Another study showed a 28% lower mortality risk for heart disease and 40% lower risk for all cancers for vegetarians [190]. Vegetarian diets reduce the risk of colorectal cancer [191-193], heart disease, stroke, diabetes, renal failure, and cancers of the breast, colon, and prostate [194], and processed meat (hot dogs, sausage, bacon, cold cuts) are particularly risky for cancer and heart disease [94, 193]. Another problem connected to meat in the diet is the widespread use of hormones and antibiotics either to increase growth rates of food animals or help prevent the sicknesses that come from the disease-laden conditions found on factory farms [195-196]. Drug-resistant bacteria "sicken 2 million Americans each year and kill 23,000 others" [195, 197] as well as 25,000 Europeans [196]. Some bacteria have been found that are resistant to all known antibiotics, and drug-resistant pathogens are considered a national security threat as serious as terrorism [196].

So, it appears that saving animals by not eating them saves many people's lives. What about climate change? Analyses carried out for greenhouse gas emissions around the world show that animal agriculture may be the single biggest source of climate-changing greenhouse gases, more than transportation, comparable to any other source of emissions [94, 198], as was described earlier. It is also one of the top contributors to environmental degradation, including deforestation, desertification, pollution, water usage and contamination, and energy usage [94]. Deforestation to clear land for animal agriculture, the generation of vast amounts of gas-creating manure, transportation related to animal agriculture, production of pesticides, and the CO_2 being generated by the breathing of 70 billion animals all contribute to climate change.

Whether animal agriculture contributes a fifth of total greenhouse gas emissions or half as some calculate, it represents a major way of fighting climate change. By cutting back on meat consumption, there is less CH_4 from enteric digestion, less deforestation, less manure and N_2O production, less land used for agriculture instead of wildlife habitat, less wasted water, less pollution, less pesticides, less chemicals, and less CO_2 exhalation from billions of animals. Reducing meat consumption, especially in a world with rising human population and rising world-wide demand for food, could be a powerful technique for helping to save the planet and all its inhabitants from the emerging global warming/climate change crisis [82, 93-98, 177]. It would also help feed the growing global human population by feeding crops to people instead of using 80 to 90% of crops to produce meat. It takes nearly 25 pounds of crops to produce one pound of beef, 9 pounds to produce a pound of pork, and 4.5 pounds to produce a pound of chicken [199]. It also takes 5 pounds of wild-caught fish to obtain 1 pound of farmed salmon. If the world demand for meat continues, more land will have to be cleared for animal agriculture, with accompanying devastation to wildlife habitat, loss of biodiversity so important for all of us, and damage to human health and well-being.

Energy

There is another benefit of fighting climate change that would help people and, indirectly, help animals. Fighting climate change brings a myriad of opportunities for progress: jobs, innovation, inventions, infrastructure development, and wellbeing. Renewable energy is well-known to everyone: solar panels, solar farms, wind turbines and wind farms, architectural design and building innovations, and geothermal power generation. The solar sector alone has already created hundreds of thousands of jobs for panel manufacturers and installers, as has the manufacture and set-up of wind turbines.

The renewable energy sector and all its associated technologies and innovations would bring hundreds of thousands if not millions of jobs, and every kilowatt generated this way is another kilowatt without greenhouse gas emissions. Electricity from benign sources like these can be used to generate a fuel, hydrogen, from fresh or sea water, and when the hydrogen is burned, the end product is only water. Creating a more efficient electrical transmission grid would reduce electrical loss and reduce the need for more power stations, whether coal, gas, oil, or renewable, to meet the increasing demand of an increasing world population. Home design and retrofit are already undergoing a job-creating and energy-reducing expansion as homes are better sealed against winter heat loss and summer cooling needs. The increasing frequency and intensity of heat waves and the higher average daily temperature will increase demand for efficient air conditioning, fueling innovations in building design and research in cooling technologies, especially valuable in developing countries and the Middle East where temperature extremes are becoming more life-threatening every year.

Research, development, design, jobs, innovation, infrastructure improvement, all are opportunities that come from fighting climate change, while saving the planet, and saving the people and animals on it.

NATURAL OR MAN-MADE

Since it's been established that climate change is happening, the question is whether it is all the result of natural causes or whether man has contributed to it. There have been many ice ages and warming periods over the last millions of years, roughly one every 80-90,000 years, so perhaps the change going on today is just more of the same, so the deniers would say. The evidence we need to find the answer is abundantly available: ice cores, tree rings, ocean and fresh water sediments, geological formations, coral reefs, and other evidence that can be used to put the story together. Ice cores in particular give data going back a million years or more. Nature gives us a library of evidence to call upon.

The Earth, its oceans, mountains, and land areas, represents a huge "thermal mass." It takes enormous amounts of energy to make significant changes in the average temperature. Even a 1°C temperature rise takes gigantic amounts of energy and takes place by natural occurrence over a minimum of 1000-2000 years. But in just the last 50–100 years, the temperature has risen 1°C [200-201], a rate at least 10 times what could be explained by natural causes [200, 202-203]. Most of this warming has occurred in the last 35 years, and 16 of the 17 warmest years on record have occurred since 2001 [200]. The hottest summer of the last 20 years, when heat waves killed thousands of people, could become the new average [202]. Over the next century, if emissions remain on their present course, the temperature increase is expected to be 2 to 5°C [202-204], a rate 50-100 times what could be explained naturally [204-205]. In the past, a 2 to 5°C temperature rise did indeed take place naturally, several times in fact, but it took 20,000 years for that change to occur. Now it's happening in a few decades [204-205].

What about greenhouse gases? Ice core data shows that CO_2 in the atmosphere has fluctuated between 180 and 280 parts per million for hundreds of thousands of years [206] and has hovered between 270 and 280 ppm for the last 2000 years. The global temperature in those years correlated very closely with the CO_2 concentration, rising and falling in tandem. Just in the last 100 years, the CO_2 concentration has shot up to 400 ppm [206], and is on a steep incline, expected to rise to 500 ppm by 2100 as the temperature climbs 2-5°C [202, 206-208]. Such a rise in CO_2 concentration hasn't occurred for millions of years.

At the same time that the earth's temperature has risen, the solar intensity, the incoming energy that fits into the energy balance equation, has fluctuated by less than 0.08% and *is actually less now than when the global temperature started its rise* over the last 40 years [206]. Global temperature and solar intensity are actually moving in the opposite direction, and therefore sunlight changes had nothing to do with the temperature rise.

Finally, the oceans have been absorbing most of the excess heat energy, and also absorbing much of the excess man-made CO_2 emissions, making the oceans more acidic than the average conditions that existed over millions of years when the coral reefs were building and flourishing. As a result, the oceans have become 30% more acidic over just the past century, a condition not seen for 300 million years, and the rate of acidification is accelerating as even more emissions pour into the atmosphere and into the ocean [209].

Bottom line: yes, climate change can happen naturally, but at a rate 100x slower than is happening now. It's the *rate* at which climate change is happening that proves it's largely man-made, not naturally-occurring. Only human influence can cause this degree of change in such a short time, and the rate of change is accelerating, to the great peril of humans and animals alike.

SUMMARY

Here is a short summary of the consequences humankind, wildlife, and the planet are facing due to the present and developing climate change. This is the path we are on, and most of it could be avoided, if we cared enough and were wise enough to act in time.

Periods of extreme temperatures are expected, life threatening in parts of the globe, with summers that are longer and hotter and require increased energy usage for air conditioning even for survival, let alone comfort.

Coastal cities could become flooded with storm surges, high tides, and rising sea levels requiring new forms of infrastructure: levies, dams, sea walls, drainage systems. High economic costs can be expected.

Hotter, dryer weather and earlier snow melt will take place that increase the number and extent of wildfires, with fire seasons that start earlier and last longer into the year, burning more acreage. Smoke from wildfires will cause air quality deterioration and human health problems.

More frequent and intense precipitation will occur, due to increased moisture in the atmosphere, with accompanying mud slides and floods.

Droughts may happen in some locations and floods in others as weather patterns change. Agriculture can be considerably damaged by either floods or droughts.

Climate change will continue to escalate extreme weather events, increasing their frequency and intensity, making heat waves hotter, droughts drier, tropical storms wetter, hurricane winds stronger.

Melting of sea and land ice will increase in the Arctic and Antarctic as well as Greenland, and hundreds of glaciers all across the world may disappear. In some locations, glaciers and snow used for fresh water supply will be reduced, endangering these communities with future water shortages.

Increased sediment pollution and contamination can take place from heavy downpours. Sea water can also contaminate fresh water supplies when storm surges and rising sea levels move inland.

Deserts may expand in size as increasing temperatures and increasing droughts affect the areas around them. Some lakes have either greatly reduced in size or even dried up completely.

Sea level rise could be enough to cause major damage to many cities and totally inundate low-lying island nations, leading to future refugee crises and possible conflicts over resources.

Melting of permafrost will continue in the north including Alaska, causing infrastructure deterioration including road and building destruction.

The acidity of the oceans is expected to increase as they absorb more CO_2, making it more difficult for some marine organisms such as clams, oysters, crabs, and corals to survive. Warmer ocean temperatures and acidity will continue to destroy coral reefs around the globe.

The destruction of forests with millions of old growth trees may continue due to insect pests such as the pine bark beetle, whose populations have exploded due to warmer weather.

Wildlife will be increasingly harmed by habitat destruction, increase in disease-carrying agents, temperature increases, droughts, floods, and wildfires. The rate of species extinction has increased due to climate change. Mass extinction of species is expected.

Human and animal health could be harmed in many ways. Intense heat waves have led to many fatalities. Smoke, ozone, and smog in general have decreased air quality and led to respiratory disease problems. Mosquitoes, ticks, and other disease-carrying agents have increased in population and expanded their ranges. Bacteria and other pathogens responsible for many diseases have also increased in density and range. Infectious diseases are on the increase.

More refugee crises can be expected as portions of land area becomes increasingly uninhabitable. Local, regional wars over diminishing resources are also likely unless the world's population develops a paradigm for sharing what will become dwindling resources.

CHAPTER 6

Nuclear War

Are these the shadows of things that Will be, or are they the shadows of things that May be only? Men's courses will foreshadow certain ends, to which, if persevered in, they must lead. But if the courses be departed from, the ends will change.
Say it is thus with what you show me!

- Ebenezer Scrooge, Charles Dickens, <u>A Christmas Carol</u>

This will be a short chapter. We all know how devastating for all life an exchange of nuclear weapons would be, but, for completeness in showing how helping animals is helping people, and helping people is helping animals, it's worth taking a look at this issue.

What are the consequences of a nuclear war? Scientists and governments have been studying this question since the first atomic bomb was dropped on Hiroshima. When hydrogen bombs were developed a few years later, with hundreds and even thousands of times the strength of the Hiroshima and Nagasaki bombs, the question became far more urgent. A single nuclear weapon can utterly destroy a city the size of New York, London, or Moscow, and the world's nuclear arsenal includes thousands of such weapons. There are now nine nuclear countries, with double that number considering joining the club. The U.S. and Russia have around 7000 warheads each, with France, China, and the UK having 200-300 each [1]. India, Pakistan, and Israel have around a hundred, and North Korea around 10, while other countries such as Iran have been trying to join the club. Altogether, there are about 15,000 nuclear weapons in the world arsenal, enough to wipe out all life on Earth many times over. Several decades ago, there were more than 70,000 nuclear warheads, with the largest number centered in Russia [2]. Fortunately that number has been reduced.

Decades ago, scientists showed that a major nuclear war could result in what was called a "nuclear winter" where sunlight would be partially blocked, temperatures would fall, and growing food would be increasingly difficult [3-4]. The concept was ridiculed at the time, mostly by those with vested interests in denying these possibilities. Since then, climate models that predict the effects of nuclear war have become very advanced and have verified the nuclear winter concept, if anything showing that the effects would be worse than originally predicted. Some of the predictions have also been tested and verified indirectly by examining historical events such as volcanic eruptions, particularly the massive eruption of Mount Tambora in Indonesia in 1815, and the nuclear accident in Chernobyl in 1986. After the eruption, which put large amounts of soot high in the atmosphere where winds distributed it around the globe, 1816 became known as the "Year Without A Summer" with average temperatures lowered by a few degrees and crop-killing frosts every month. Crop prices skyrocketed and livestock prices plummeted. In Europe the weather was so cold that the stock market collapsed and widespread famine occurred [5].

Around Chernobyl, the release of radioactive cesium 137 rendered 1100 sq. miles uninhabitable and contaminated 5 million acres of cropland, with major damage to wildlife including sterility, cancer, and partial blindness [4, 6-7].

Mathematical simulations have been carried out for both an all-out nuclear war involving hundreds or thousands of high yield warheads exchanged by opposing sides and for a limited war involving 50 to 100 Hiroshima-sized bombs. A hypothetical regional war between India and Pakistan with detonations of 50 atomic bombs from each side was taken as an example.

The consequences of the smaller, regional war were examined first. In addition to tens of millions of immediate deaths from blast effects, thermal radiation, and firestorms, the *global* climate and agriculture would be so damaged that more than a billion people would be at risk of starvation [8-10]. The simulation indicates that around 6 million metric tons of smoke and soot would be injected into the atmosphere, reducing the incoming sunlight by at least 10%, reducing rainfall, and producing a drop in temperature, shortening the growing season, threatening agriculture worldwide and causing food shortages, especially to the world's poorest people. Infectious disease epidemics would occur in multiple locations and local wars over scarce resources could likely become common [2, 8-10]. Again according to the simulation models, corn and soybean production would be reduced by 10-20% as far away as the U.S. and rice by 20% in China. Most of these predictions, except for local wars, were consistent with the observed effects of the Tambora volcanic eruption. The infectious disease predictions that can accompany famine were supported by, for example, the Great Bengal Famine of 1943 in which 2-4 million Indians died, mostly of starvation, but which also caused outbreaks of cholera, malaria, smallpox, and dysentery [10-11].

Large quantities of industrial chemicals and pollutants could also be lifted into the atmosphere by blast damage and fires in factories and cities [12-13]. Equally damaging, the protective ozone layer in the upper atmosphere would be significantly reduced, allowing dangerous UV radiation to reach the Earth's surface, leading to increases in skin cancer, crop damage, and eye damage [3, 9, 14]. This excess UV radiation would also have a highly damaging impact on plants and animals, with aquatic ecosystems especially at risk. Amphibians are particularly susceptible to UV, and ocean plankton could be destroyed, the plankton which is the starting point of the entire marine food chain [14] and the major contributor to the world's oxygen in the atmosphere.

All this would occur from a limited, regional nuclear exchange involving only 100 low yield bombs, equivalent to a total of about 1-2 Megatons of TNT. In modern times, most nuclear devices are tens to hundreds to thousands of times more powerful than the Hiroshima bomb, and both the USA and Russia have monstrous nuclear devices with 40 to 50 Megatons of TNT explosive power [15], 2500 times the strength of the Hiroshima/Nagasaki bombs. Simulations of a nuclear exchange between nuclear superpowers like Russia and the U.S. predict the annihilation of virtually all life on Earth, including the extinction of the human species and almost all animal species.

A large scale nuclear war between the U.S. and Russia, involving even a portion of their arsenals, could inject 150 million metric tons of smoke, soot, and chemicals into the atmosphere where it would be distributed around the globe by the atmospheric winds and remain for many years [3, 13]. Sunlight would be blocked by 35% in the southern hemisphere and 70% in the northern, creating nighttime conditions at noontime. Rainfall could be cut in half, creating the possibility of clean drinking water shortages and water riots. Global temperatures could drop by 10 to 20°C (18-36°F) causing multiple recurring frosts and virtually eliminating the global growing season, causing mass starvation. Radioactive fallout would be widespread, and the aforementioned infectious disease epidemics would be multiplied many times over. Health care would be virtually non-existent as the world ran out of medicines and antibiotics to fight diseases. The ozone layer would be nearly gone, allowing dangerous UV exposure for any human or animal outdoors. Local wars over reduced or rapidly vanishing resources would be sure to take place, for the short time that humans continued to exist.

In short, no one could win a large scale nuclear war. The Earth would become uninhabitable for all species except for a few cave-dwelling animals or insect species that are less susceptible to radioactivity and some deep-dwelling ocean species, assuming the deep ocean has enough thermal mass to sustain them for the decades it would take for the sunlight to return and re-warm the planet. Even those deep-dwelling species would probably starve as there would be no significant life left on land or the upper levels of the ocean to supply them with food.

Only humans can prevent these nuclear holocausts from happening, whether regional or large scale, but all the planet's animal, plant, and invertebrate life, innocent of causing the destruction, would suffer and die along with their human companions. It's amazing that the nuclear weapons industry fought against reducing nuclear weapons arsenals and the nuclear winter scenario in the early days, and even today, government and military leaders virtually ignore the scientific findings about what the lasting effects of a nuclear exchange would be [3-4]. It's almost comical that scientists and others become highly excited at the thought of finding some tiny life form on another planet, Mars, or the moon, while mankind does its best to drive exquisite species great and small to extinction on our own planet, by possible nuclear confrontation, climate change, wildlife policies, or just simply not caring.

Helping animals by not destroying them in a nuclear holocaust would certainly be helping people. To put it more strongly, saving the lives of countless animals by preventing their destruction from a nuclear warfare exchange has definite benefits for humans.

CHAPTER 7
A Few Suggestions

Animals are the ultimate art of the universe. They are at once functional, appealing, mysterious, and inspiring.
- The tragedy of today is that we are embarked upon a campaign to destroy these masterpieces of creation. One by one we crowd them toward extinction with gun, spray, saw, match, bulldozer, bomb, markets, indifference, and procrastination.

- Dr. Alfred G. Etter

In this ongoing travesty against nature, this ongoing war on wildlife, the good news is that helping animals brings solid benefits to humans as well as the animals. It was described in previous chapters how animals can be helped. In the chapters on hunting, it was obvious that cutting back on hunting would save some of those 100 human lives lost and 1000+ accidents that take place. An even larger benefit would come from the reduced deer-car collisions that take even more lives and cost billions of dollars in car repair alone, in addition to the doctor's bills, hospital stays, and possible funeral costs. A more subtle benefit would come in the form of less Lyme disease and all the serious co-infections that can come from tick and mosquito bites: babesia, bartonella, anaplasma, Q-fever, and more. An even more subtle benefit would come from the ability to walk in a Wildlife Refuge or Park without the fear of being hit by a stray bullet or suffer the agony of stepping into a leghold trap.

Hundreds of park rangers in Africa would be alive today and hundreds more in the future if the poaching problem would be tackled in a serious way, with the sophistication of modern weapons and modern detection techniques, as well as banning ivory use and sales worldwide. Many species would not go extinct if humans made their survival a priority instead of turning a blind eye to their decimation in the bushmeat, shark finning, whaling, and sport hunting endeavors. Educating the Asian cultures that rhino horn is the same as fingernail material and has no medicinal value, and that tiger or lion organs and bones do not boost men's virility, or that many medicines exist with far better value than bear bile, would help both people and animals.

Mankind would also benefit from the recognition that all species on Earth have a part to play in a diverse and healthy ecosystem, and interfering with Nature's plan can have unforeseen negative consequences. Even the smallest invertebrate has a part to play and humans benefit in countless ways, most unseen and unknown, by all the gifts bestowed to us by the gigantic, intricately functioning biome that is the world-wide ecosystem of the planet. Believe it or not, the frogs, worms, bees, bats, birds, many types of insects, even the beneficial bacteria, and countless others, all play a role in keeping us all healthy.

Human beings get much of their protein intake from eating animals: farm animals and seafood. This is a risk that we are willing to take, that we won't be the ones to suffer from the many diseases that can potentially accompany this type of diet: the cardiovascular diseases, cerebrovascular diseases, diabetes, and so forth, along with the antibiotics, growth hormones, and chemicals that also accompany modern agriculture, and the poisons and chemicals that can be present with seafood. Fortunately for us, the consequences of animal diets may only show up late in life, or are rare such as food poisoning from eating fish. People don't *need* animal protein to survive, particularly red and processed meat, as the medical data comparing vegetarian and vegan diets with non-vegetarian shows. Some uncommon medical conditions may suggest the need for animal protein, but for most of us, the data shows we'd be healthier without it.

What can we do about climate change? Of course, there's the obvious, like driving more fuel-efficient cars, using energy-efficient appliances and lighting, turning the thermostats up a degree in the summer and down a degree in the winter. Planting a tree or two has real benefit; trees are an important carbon sink and help supply oxygen. If mankind were to restore the forests, a major benefit would ensue. Deforestation, whether by design to make more room for cattle or through disease like the expanding pine bark beetle, is a major source of carbon dioxide emission into the atmosphere and prevents the huge benefit of forests acting as carbon sinks.

How about renewable energy? Shifting from fossil fuels to benign energy sources such as solar panels, wind turbines, biomass, geothermal, solar hot water, maximizing hydroelectric, and improving energy transmission efficiency, would reduce our reliance on fossil fuels. Transforming our energy paradigm would create millions of jobs, may create whole new industries, and they would be jobs that still require human hands rather than strictly relying on robots. What about nuclear power? It would certainly reduce fossil fuel use, but until we are willing to spend the money to make it safer and until we solve the spent fuel storage problem, it will remain a risk that other forms of renewable energy don't have.

Animal agriculture is a major contributor to climate change, some say *the* major contributor, by deforestation, methane and CO_2 emissions, pesticide production, and energy and water use. We humans have been brought up to think we require animal protein, yet people along with many animal species like horses and elephants do just fine without it. If animal agriculture in all its aspects really is a major source of greenhouse gases, a major way of fighting climate change is to eat less meat. Even making a day a week a meatless day would have beneficial effects, and two days a week would be even better.

So many human lives and animal lives would be saved by fighting climate change! From heat waves and droughts, floods, storm surges, wildfires, food shortages, water shortages, local wars over resources, refugee crises, expanding and more virulent diseases, all life would benefit from stopping this runaway train that threatens us now and threatens future generations much more. Eat less meat, drive smaller cars, use energy efficient lighting and appliances, plant more trees; these are things we all can do instead of relying solely on government officials to solve the problem for us.

The benefits to animals and humans by preventing nuclear war are obvious. A major nuclear exchange could be a civilization-level and mass extinction-level event. Even a limited, smaller war has global consequences. Will we humans stop our saber-rattling and learn to get along instead of expanding weapon size, number, and potency? Good question!

Short list of activities to help people and wildlife. (Not in any order of priority. Some of these all of us can do in our own lives, others we could encourage through government action.)

Eliminate buck laws that promote deer populations.
Eliminate wildlife killing contests.

Outlaw canned hunts and wildlife (pheasant) stocking
    programs.
(Dare I say, wish for, prevent, hunting altogether?)
Outlaw leghold traps.
Bring back apex predators to re-establish nature's balance.
Eliminate "Wildlife Services" (United States).
Outlaw the M44 cartridge and other poisons used to kill
    wildlife.
Re-establish Wildlife Refuges as true wildlife sanctuaries.
Eliminate the wildlife "trade," which helps to fuel terrorism.
Fight against tiger, lion, and leopard "farms."
Oppose overfishing, whaling, and shark finning.
Promote humane education in schools, teaching the benefits
    of wildlife and ecology.
Ban ivory and ivory products worldwide. Don't buy ivory.
Fight poaching and the bushmeat trade.
Preserve rainforests.
Support endangered species everywhere. Fight the forces of
    extinction.
Keep your cats indoors. (Stray cats kill a lot of wildlife)
Plant a few trees.
Plant forests.
Encourage your government to fight climate change.
Choose and drive more fuel efficient vehicles.
Drive less miles. Combine trips. Carpool.
Use public transportation where possible.
Prevent methane leaks from oil, coal, and gas production.
Promote renewable energy: install solar panels, solar water
    heaters.
Renewable electrical energy + electric cars = much less CO_2
    from fossil fuels.
Use efficient lighting, especially LED (Light Emitting
    Diode) lighting.
Buy the most efficient appliances.

Turn the thermostat down a degree in winter and up a degree in summer. Even more is better.

Increase home energy efficiency: install insulation, prevent air drafts, use efficient furnaces.

Use antibiotics sparingly.

Eat organic food.

Eat less meat.

Eat less seafood too.

Use less plastic. Recycle.

These are just a few things we can do, in our personal lives or through activism, legislation, and even by setting an example. Much can be done to improve the outlook for people and animals, if we only cared enough.

EPILOGUE

What is man without the beasts? If the beasts were gone, men would die from a great loneliness of spirit. For whatever happens to the beasts, soon happens to man. All things are connected. Whatever befalls the earth befalls the sons of the earth. Man did not weave the web of life, he is merely a strand in it. Whatever he does to the web, he does to himself.

- Chief Seattle (Suquamish Tribe of native Americans)

Webster's Dictionary lists several definitions of *war* and *warfare*: "active hostility or contention, conflict," and "conflict, especially when vicious and unrelenting." It seems that poaching, hunting, predator killing contests, extinction of species, government-sanctioned wildlife killing, canned hunts where wildlife has no chance, trapping, and the metamorphosis of so-called national wildlife refuges and wilderness areas into priority hunting and trapping preserves, fit these definitions of a war on wildlife quite well.

It's not that benefiting animals is in conflict with helping people - quite the contrary. Helping animals would save human lives and human suffering in so many ways. If Wilson and the Erlichs are right [Chap. 2, 29-30], the long term survival of human beings, life as we know it, are dependent upon understanding and respect for nature and the value of biodiversity, the value of the countless species that make up the Earth and keep it healthy. The survival of any species, including our own, depends in no small measure on the survival of the others.

It seems clear by now that homo sapiens, by fighting nature instead of embracing it, by placing short-term economic gains over long term welfare especially for future generations, by damaging and destroying the very life forms that a balanced Earth depends on, by killing his fellow creatures either for sport or for having the audacity to interfere with man's chosen activities, is sowing the seeds of his own destruction. These are cumulative, long term effects; the damage will be felt mostly by future generations, just as the effects of climate change will be visited on future generations. The term "temporal bias" has been used to describe the feelings of present living individuals toward future generations; people care less and less about the future as it becomes more distant. To leave future generations a healthy Earth, those of us living today have to love and care for individuals that we'll never know, some of whom will be our own descendants, allowing them to inherit a healthy planet instead of a deteriorating Earth and an escalating set of problems. We have to love the Earth itself.

Since there is a well-established link between cruelty to animals and violence to humans, especially domestic violence, one has to wonder about the mindset of the people willing to inflict such pain and suffering on living beings whose only crime was not to be born human. For children especially, these activities, in which many children spectate or even participate, can't be good learning experiences for instilling compassion, tolerance, and respect for life and for nature.

And there are always the more subtle forms of cruelty where

many people may not even recognize the cruelty at all: elephants chained by the leg for 23 hours a day while they're not performing; chimpanzees and bears confined to tiny cages in roadside zoos; lion, leopard, and tiger farms in Asia where these magnificent great cats are confined in enclosures all their lives on their way to being shot for their body parts; dolphins and orcas forced to swim endlessly in small pools at aquariums or ocean-world parks instead of the vast oceans that would otherwise be available to them. Despair, endless boredom, and suffering are not the exclusive prerogative of homo sapiens.

Is this assessment too pessimistic, too dramatic? In this book, which represents just a subset of the ways helping animals means helping people, significant damage to the Earth is being perpetrated just by the war on wildlife and its affiliates: poaching, species extinction, connections to terrorism, short term economic gain, wildlife practices, and ignorance of nature. Add all the other mistakes homo sapiens commits daily and seems to be oblivious to, and the future could be bleak in the long run. People seem to care less about wildlife as time goes on, not more, and we are all wrapped up in our daily lives and the non-ending media coverage of wars, suicide bombings, terrorism, political differences, religious animosity, and intolerance between races and religions. Perhaps we don't have enough energy left to care about anything else. All the other homo species went extinct; are we too arrogant to think that it could never happen to us? Are we already exceeding the carrying capacity of our species, as some have suggested, with human population growth and its ever increasing need for resources? Are we destroying the very non-human life that could help us survive? Will we realize how precious a gift we have been given before it's too late? Will we ever realize that by destroying other species, by our indifference to their suffering and death, we are destroying ourselves? It doesn't have to be this way!

Champions of wildlife are few and far between and have little political clout, while exploiters are largely ruthless and often have governments behind them. The world is so wrapped up in local and global racial, religious, and political conflict that it cares very little for non-human life (and often human life as well) amid the chaos. Like climate change, the damage will be measured in decades, and just as with climate change, it will likely be irreversible.

Assuming there's no nuclear war. Then all bets are off.

REFERENCES

CHAPTER 1.

[1]. L. Foote et al, "The Rational Hunter, Animal Pain, Animal Suffering," http://rationalhunter.typepad.com/rational_hunter/2004/07/animal_pain_ani.html

[2]. R. Baker, <u>The American Hunting Myth</u>, Vantage Press, New York, 1985.

[3]. S. Ditchkoff et al, "Wounding Rates of White Tailed Deer With Traditional Archery Equipment," Proc. Annual Conf. Southeastern Assoc. of Fish and Wildlife Agencies, 1998. Eleven deer out of 22 that were shot were wounded and not recovered by the hunter.

[4]. V. Spencer, "Reducing Wounding Losses," S. Dakota Dept. of Game, Fish, and Parks, 25 July 2013.

[5]. Reference 2, page 69.

[6]. M. Scully, "Hunting for Fun and 'Charity'," New York Times, 4/27/99.

[7]. The Fund For Animals, Newsletter, Volume 28, Number 2. Fall 1995.

[8]. M. Kayser, "The Good, The Bad, and the Ugly About Coyote Calling Contests," GrandViewOutdoors.com, Feb. 4, 2013.

[9]. D. Flores, "Stop Killing Coyotes," New York Times 8/11/2016; pg. A23.

[10]. W. Elliot, "The Competition Borders on Intense in Rabbit Hunt Fundraiser," The Buffalo News, Jan. 23. 2016.

[11]. northamericanhuntingcompetition.com

[12]. "Gun Lobby Helps States Train Young Hunters," New York Times 11/17/1999, pg. A1.

[13]. "Pheasant Stocking, an Inefficient Management Tool," phesasntsforever.org

[14]. J. Walls and A. Pearson, "Cheney Faces Heat for "Canned Hunt,"msnbc.com,12/18/2003. www.today.com/id/3675813/ns/today.../t/cheney-faces-heat-canned-hunt/

[15]. Darryl Fears, "USDA's Wildlife Services Killed 4 Million Animals in 2013; Seen as an Overstep By Some," Washington Post, June 7, 2014.

[16]. https://www.aphis.usda.gov/wildlife_damage/prog_data/2014/G/Tables/Table_G-2_Euth-Killed.pdf

[17]. P. Clark, "The Toll Taken By Wildlife Services (USDA)," Washington Post, June 6, 2014.

[18]. T. Knudson, "Wildlife Services' Methods Leave a Trail of Animal Death," *Sacramento Bee*, April 30, 2012.

[19]. Cristina Corbin, "Animal Torture, Abuse Called a 'Regular Practice" Within Federal Wildlife Agency," FoxNews.com, March 12, 2013.

[20]. Cristina Corbin, "Federal Agency Gives Few Answers on Months-Long Probe of Alleged Animal Cruelty," FoxNews.com, June 12, 2013.

[21]. T. Knudson, "The Killing Agency: Wildlife Services' Brutal Methods Leave a Trail of Animal Death," *Sacramento Bee*, April 28, 2012.

[22]. "EXPOSED; The USDA's Secret War on Wildlife," video available on www.predatordefense.com

[23]. T. Knudson, "Wildlife Services' Deadly Force Opens Pandora's Box of Environmental Problem," *Sacramento Bee*, April 30, 2012.

[24]. T. Knudson, "Investigation into Wildlife Services Methods Sought By Four House Members," *Sacramento Bee*, June 25, 2012. Darryl Fears, "Petition Targets 'Rogue' Killings By Wildlife Services," Washington Post Dec. 15, 2013.

[25]. Michael Mares, President, American Society of Mammologists, quoted in reference [18].

[26]. "The Controversy Over the Federal Government's Predator Control Program," HealthNewsDigest.com, Nov. 17, 2012.

[27]. "America's Wildlife Body Count," New York Times 9/18/2016; pg 12.

[28]. World Wildlife Fund, www.worldlife.org/threats/illegal-wildlife-trade

[29]. B. Rohr, "WWF Statement on Southern Africa's Alarming Rhino Poaching Rates Reported for 2015," Jan.21, 2016. https://www.worldwildlife.org/press-releases/wwf-statement-on-southern-africa-s-alarming-rhino-poaching-rates-reported-for-2015

[30]. K. Gibson, "Proposed U.S. Ban on Ivory Trade Faces Powerful Foe," CBS News MoneyWatch, July 31, 2015.

[31]. "U.S. House Passes 'Global Anti-Poaching Act' That Puts Wildlife Trafficking on Par With Drug and Weapons Trafficking", Nov.3,2015; http://news.mongabay.com/2015/11/u-s-house-passes-global-anti-poaching-act-that-puts-wildlife-trafficking-on-par-with-drug-and-weapons-trafficking/

[32]. "The Hard Truth: How Hong Kong's Ivory Trade is Fueling Africa's Elephant Poaching Crisis," World Wildlife Fund, 2015 Report; http://assets.worldwildlife.org/publications/816/files/original/wwf_ivorytrade_eng_eversion.pdf?1442844784&_ga=1.46437156.221690821.1432506688

[33]. S. Guynup, "Illegal Tiger Trade: Why Tigers Are Walking Gold," Cat Watch, Feb 12, 2014. http://voices.nationalgeographic.com/2014/02/12/illegal-tiger-trade-why-tigers-are-walking-gold/

[34]. "U.S. Spends Millions to Expand the Fight Against Poachers in South Africa," New York Times 12/23/2015.

[35]. P. Baker and J. Smith, "Obama Administration Targets Trade in African Elephant Ivory," New York Times July 25, 2015.

[36]. "Tigers in Crisis," http://www.tigersincrisis.com/trade_tigers.htm

[37]. L. Dattaro, "Obama Has Proposed a Ban on Almost All Ivory Sales in the United States," news.vice.com, July 27, 2015 https://news.vice.com/article/obama-has-proposed-a-ban-on-almost-all-ivory-sales-in-the-united-states

[38]. R. Schara, "Cecil the Lion, Ivory Smuggling, and Outrage," Star Tribune, Feb. 1, 2016 http://www.startribune.com/cecil-the-lion-ivory-smuggling-and-outrage/367290621/

[39]. "Poaching: The Statistics," https://www.savetherhino.org/rhino_info/poaching_statistics

[40]. "As Few as 3200 Tigers Left," World Wildlife Fund, http://www.savetigersnow.org/problem. "Animal Farms," New York Times 6/6/2017; pg. D1

[41]. http://www.awf.org/campaigns/poaching-infographic/ African Wildlife Foundation

[42]. http://www.fws.gov/international/wildlife-without-borders/global-program/bushmeat.html_

[43]. R. Meador, "Africa's Lions Are in Steep Decline, and the Problem is Bigger than Poaching," https://www.minnpost.com/earth-journal/2015/07/africas-lions-are-steep-decline-and-problem-bigger-poaching

[44]. "Threats To Gorillas. Hunting and Poaching." http://wwf.panda.org/what_we_do/endangered_species/great_apes/gorillas/threats/

[45]. M. Goodavage, "Gorilla Poaching: The Sad, Savage Reality," Oct. 18, 2012. http://www.takepart.com/article/2012/10/17/baby-gorilla-poaching

[46]. "Elephants Face Extinction if Beijing Does Not Ban Ivory Trade: China Accounts for Nearly Half of the 40,000 Killed Every Year for their Tusks," June 17, 2013; http://www.dailymail.co.uk/news/article-2343137/Elephants-face-extinction-Beijing-does-ban-ivory-trade-China-accounts-nearly-half-40-000-killed-year-tusks.html

[47]. C. Russo, 3/8/2013, "The Elephant Massacre at Bouba Ndjida: One Year Later," http://www.huffingtonpost.com/christina-russo/the-elephant-massacre-at_b_2838828.html

[48]. H. Gold, "Hong Kong Says It's Going to Ban Its Trade in Ivory and Elephant Trophies," https://news.vice.com/article/hong-kong-says-its-going-to-ban-its-trade-in-ivory-and-elephant-trophies January 16, 2016

[49]. J. Griffiths and I. Watson, "Dark Heart of the Ivory Trade No More: Hong Kong to Phase Out Ivory Sales," http://www.cnn.com/2016/01/13/asia/hong-kong-ivory-trade/ Jan 13, 2016.

[50]. J.F. Smith, "U.S. Bans Commercial Trade of African Elephant Ivory," New York Times, June 2, 2016; http://www.nytimes.com/2016/06/03/world/africa/elephant-ivory-ban.html?_r=0

[51]. "China Banning Ivory, Thrilling Nature Groups," New York Times 12/31/2016; pg.A1.

[52]. "India's Tiger Poaching Crisis," http://www.wpsi-india.org/tiger/poaching_crisis.php

[53]. "Thailand Tiger Temple: Forty Dead Cubs Found in Freezer." BBC News, June 1, 2016; http://www.bbc.com/news/world-asia-36424091

[54]. Gorillas World, http://www.gorillas-world.com/gorilla-hunting/

[55]. E. Conway-Smith, "Ebola Has Killed a Third of the World's Chimpanzees and Gorillas," The Telegraph, Jan. 22, 2015. http://www.telegraph.co.uk/

[56]. J. Randerson, "Congolese Chimpanzees Face New Wave of Killing for Bushmeat," http://www.theguardian.com/environment/2010/sep/07/congo-chimpanzees-bushmeat, Sept. 7, 2010.

[57]. http://www.janegoodall.ca/chimps-issues-bushmeat-crisis.php

[58]. http://www.koko.org/

[59]. C. Fairclough, "Shark Finning: Sharks Turned Prey," http://ocean.si.edu/ocean-news/shark-finning-sharks-turned-prey

[60]. J. McCurry, "Shark Fishing in Japan – a Messy, Blood-spattered Business," http://www.theguardian.com/environment/2011/feb/11/shark-fishing-in-japan

[61]. The International Whaling Commission, https://iwc.int/home

[62]. R. Halter, "Japanese Whale Poachers Masquerade Behind Thin Veil," June 13, 2014. http://www.huffingtonpost.com/dr-reese-halter/japanese-whale-poachers-m_b_5493548.html

[63]. WDC, Whale and Dolphin Conservation Campaigning, http://us.whales.org/wdc-in-action/whaling

[64]. "End Commercial Whaling; Renegade Whale Hunting Threatens the Survival of Endangered Species Around the World," National Resources Defense Council, http://www.nrdc.org/wildlife/whaling.asp

[65]. L. Dye, "How Whale Hunting Changed the Ocean," Sept. 25, 2003. http://abcnews.go.com/Technology/story?id=97519&page=1

[66]. P. Weihe and H.D. Joensen, "Dietary Recommendations Regarding Pilot Whale Meat and Blubber in the Faroe Islands," Int J. Circumpolar Health. 71, 18594, (2012). http://www.ncbi.nlm.nih.gov/pmc/articles/PMC3417701/

[67]. "Blue Voice,org Documents Brutal Slaughter of Dolphins in Japan and the Tie to the Dolphin Captivity Industry," http://www.bluevoice.org/news_dolphinsincite.php

[68]. Y. Wakatsuki and M. Park, "Japan Officials Defend Dolphin Hunting at Taiji Cove," http://www.cnn.com/2014/01/20/world/asia/japan-dolphin-hunt/

[69]. African Lion Preservation, "Human Poaching of the Savannah Mammals," https://africanlions.wikispaces.com/Human+Poaching+of+the+Savannah+Mammals

[70]. Endangered Species Handbook: Bears. http://www.endangeredspecieshandbook.org/trade_traditional_bears.php

[71]. "Bile Bear," https://en.wikipedia.org/wiki/Bile_bear

[72]. https://www.opencongress.org/bill/hr3480-111/bill_positions

[73]. "A Danger to Our Grizzlies," New York Times 2/24/2016; pg. A19.

[74]. C. Bush, "Captive Hunts," www.christinabush.com/canned-hunting.html

[75]. "The Unfair Chase," Captive Hunts Fact Sheet, HSUS, Aug. 17, 2012.

[76]. P. Barkham, "Canned Hunting: The Lions Bred for Slaughter," www.theguardian.com, June 3, 2013.

[77]. "Canned Hunting in the United States," www.bornfreeusa.org

[78]. "500 Canned Hunting Ranches in Texas," https://bigcatrescue.org/500-canned-hunting-ranches-in-tx Feb. 8, 2006.

[79]. M. Winikoff, "Blowing the Lid Off Canned Hunts," HSUS News, Summer 1994.

[80]. D. Norris et al, "Canned Hunts: Unfair at Any Price," The Fund for Animals, http://fund.org/library/documentViewer.

asp?ID=338&table=documents, Michigan State University, Animal Legal and Historical Center, 2002.

[81]. "Eight Shots, Three Hours, One Death: The Story of a Canned Hunt," HSUS, June 1, 2007. "Monster Pig Shows Horror of Canned Hunts," June 4, 2007.   blog.humanesociety.org/wayne/2007/06/last_week_an_11.html

[82]. R. Shearer, "Alabama's Monster Pig Hoax, One Year Later," iMedia Ethics, May 3, 2008; http://www.imediaethics.org/alabamas-monster-pig-hoax-one-year-later/

[83]. "Killing Tamed Wild Animals in Fenced Areas for Sport; Petting Clubs in Africa Supports the Canned Hunting Industry," https://bigcatrescue.org/abuse-issues/issues/canned-hunting/

[84]. "Canned Lion Hunt Footage 2102 – Hunter Kills Tame Lioness in her Enclosure in South Africa," April 3, 2013. https://www.youtube.com/watch?v=tExAqeGXRZU

[85]. K. Nowak, "Inside the Grim Lives of Africa's Captive Lions," National Geographic, July 22, 2015. http://news.national-geographic.com/2015/07/150722-lions-canned-hunting-lion-bone-trade-south-africa-blood-lions-ian-michler/

[86]. C. Ward, "The Lion Whisperer," 60 Minutes, Nov. 30, 2014.

[87]. "Blood Lions," www.msnbc.com/documentaries/bloodlions-premieres-wednesday-october-7th-msnbc; www.bloodlions.org

[88]. A. Schelling, "Lions Are Raised So People Can Shoot Them in Cages," July 31, 2015. https://www.thedodo.com/canned-hunting-cecil-lions-1274358177.html

[89]. "New U.S. Protections for African Lions Should Doom South Africa's Canned Lion Hunting Industry," Humane Society

International, Dec. 30, 2015. http://www.humanesociety.org/news/press_releases/2015/12/us-protections-south-africa-canned-lion-hunting-123015.html
http://southafrica.nuscorner.com/news/new-us-protections-for-african-lions-should-doom-south-africas-canned-lion-hunting-industry

[90]. "Lions Are Now Protected Under the Endangered Species Act," U.S. Fish and Wildlife Service, Endangered Species, http://www.fws.gov/endangered/what-we-do/lion.html

[91]. J. Tarlton and M. Green, "Explained: 'Sanctuary Cities' and What They Mean for Undocumented Immigrants," KQED News, Oct. 20, 2015. http://ww2.kqed.org/lowdown/2015/07/10/explainer-what-are-sanctuary-cities/

[92]. Born Free USA. http://www.bornfreeusa.org/facts.php?more=1&p=62

[93]. U.S. Fish and Wildlife Service, National Wildlife Refuge System. http://www.fws.gov/refuges/about/mission.html

[94]. http://www.fws.gov/refuges/RefugeUpdate/pdfs/refUp_JanFeb_2016.pdf

[95]. D. Day, The Doomsday Book of Animals, A Natural History of Vanished Species, Viking Press, New York, 1981.

[96]. R. Fuller, Lost Animals, Princeton University Press, Princeton, 2013.

[97]. "Earth's Smallest Porpoise Slips Closer to Extinction," New York Times 5/16/2016; pg. A5.

[98]. "Endangered Species on the Grill: The Black Market in Illegal Meat Flourishes in the U.S." http://www.alternet.org/story/145668/

endangered_species_on_the_grill%3A_the_black_market_in_illegal_meat_flourishes_in_the_us

CHAPTER 2.

[1]. http://www.tribstar.com/news/local_news/falls-from-tree-stands-top-hunting-accidents/article_e7689704-616d-5724-b4e3-67202a465c55.html

[2]. http://www.nssf.org/pdf/research/iir_injurystatistics2013.pdf

[3]. http://www.all-creatures.org/cash/accident-center.html. (Committee to Abolish Sport Hunting)

[4]. Merritt Clifton, http://dakotatree.tripod.com/hunting_facts.html

[5]. B. Rosner, <u>The Top 10 Lyme Disease Treatments</u>, BioMed publishing Group, Lake Tahoe, 2007.

[6]. Lyme Disease: A Growing Threat to Urban Populations, <u>Infectious Diseases in an Age of Change</u>, B. Roizman, Editor, National Academy of Sciences, National Academy Press, Washington, 1995.

[7]. Field and Stream, Aug. 18, 2015.

[8]. R. Baker, "How Hunting Contributes to Species Extinction," The CASH Courier, Spring Issue, 1997.

[9]. "A Natural Cure For Lyme Disease," New York Times 8/21/2016; pg. SR9.

[10]. The Decline of Deer Populations, http://www.deerfriendly.com/decline-of-deer-populations

[11]. M. Bensley, "Hunting is Used As a Tool to Intentionally Increase Deer Numbers," https://docs.google.com/document/d/12i1N29_c9rM_gvrkjjpv9V_5_wr3IdpU9TvXB_U3KDw/edit

[12]. "Deer vs. Car Collisions," http://www.cultureofsafety.com/driving/deer-vs-car-collisions/

[13]. Insurance Journal, Oct 24, 2012.

[14]. A. Cambronne, "Can't See the Forest for the Deer," The Wall Street Journal, 3/11/2014.

[15]. A. Rutberg, "The Science of Deer Management: An Animal Welfare Perspective," <u>The Science of Overabundance: Deer Ecology and Population Management</u>, W. McShea, H. Underwood, and J. Rappole, eds. Smithsonian Institution Press, Washington & London, 1997.

[16]. The Conservationist, New York State Department of Environmental Conservation.

[17]. C. Robbins, "Bow Season Bounty," New York Sportsman, 9/2001.

[18]. D. Arnold, Department of Natural Resources executive, Michigan, comment to Free Press, January 1, 1980.

[19]. G. Yourofsky: http://biteclubkc.wordpress.com/2009/09/18/hunters-are-the-terrorists-of-the-animal-world/

[20]. http://www.drdavidwilliams.com/9-ways-good-gut-bacteria-support-your-overall-health/#immune

[21]. "Lush Life; But As Species Vanish, What Will We Lose?" New York Times June 2, 1998, pg. G1.

[22]. http://www.iucn.org/about/work/programmes/species/ our_work/invertebrates/ International Union for Conservation of Nature.

[23]. S.R. Kellert, The Value of Life: Biological Diversity and Human Society, Island Press, Washington, 1996.

[24]. "Bugs Keep Planet Healthy Yet Get No Respect," New York Times Dec 21, 1993.

[25]. The Week, 8/20/2010, pg. 23.

[26]. http://www.pbs.org/wnet/nature/fabulous-frogs-the-worlds-most-endangered-frogs/8917/ The World's Most Endangered Frogs.

[27]. http://biodiversitygroup.org/amphibian-population-declines/ The Biodiversity Group; Amphibian Population Declines.

[28]. "Report Tallies Hidden Costs of Human Assault on Nature," New York Times April 5, 2005; pg. F2.

[29]. P. and A. Ehrlich, Extinction. The Causes and Consequences of the Disappearance of Species, Random House, New York, 1981.

[30]. E.O. Wilson, Half Earth. Our Planet's Fight For Life, Liverlight Publishing Corporation, New York, 2016.

[31]. Julia Marton-Lefevre, Director General, International Union for the Conservation of Nature, http://www.gameranger.org/news-views/media-releases/96-call-for-action-against-increasing-risk-to-game-rangers.html

[32]. "Poaching's Bloody Human Toll," New York Times 12/5/2016; pg. A23.

[33]. A. Kumar Sen, "Terrorists Slaughter African Elephants, Use Ivory to Finance Operations," The Washington Times, November 13, 2013. http://www.washingtontimes.com/news/2013/nov/13/terrorists-slaughter-african-elephants-use-ivory-t/?page=all

[34]. W. Chinhuru, "Poachers Outgun Africa's Vulnerable Wildlife Rangers," http://www.equaltimes.org/poachers-outgun-africa-s?lang=en#.VxvobvkrJD8

[35]. "Isolated Rangers in Chad Encounter Grim Cost of Protecting Wildlife," New York Times January 6, 2013; pg. 14.

[36]. L. Neme, "For Rangers on the Front Lines of Anti-Poaching Wars, Daily Trauma," National Geographic, June 27, 2014. http://news.nationalgeographic.com/news/2014/06/140627-congo-virunga-wildlife-rangers-elephants-rhinos-poaching/

[37]. The Week, March 21, 2014; pg. 11.

[38]. "Terror in Nairobi: The Full Story Behind Al-Shabaab's Mall Attack," http://www.theguardian.com/world/2013/oct/04/westgate-mall-attacks-kenya

[39]. J. Bergenas and M. Medina, "Breaking the Link Between Terrorism Funding and Poaching," The Washington Post January 31, 2014. https://www.washingtonpost.com/opinions/break-the-link-between-terrorism-funding-and-poaching/2014/01/31/6c03780e-83b5-11e3-bbe5-6a2a3141e3a9_story.html

[40]. "ISIS: New Terrorist Group Jahba East Africa Pledges Allegiance to 'Islamic State' in Somalia," http://www.independent.co.uk/news/world/africa/isis-new-terrorist-group-jahba-east-africa-pledges-allegiance-to-islamic-state-in-somalia-a6974476.html

[41]. A. Abubakar, CNN, February 13, 2016, "Boko Haram Attacks Kill at Least 30, Locals Say," http://www.cnn.com/2016/02/13/africa/boko-haram-attack-nigeria/

[42]. "Boko Haram Attack: Children Burned Alive in Nigeria," Al Jazeera, February 1, 2016; http://www.aljazeera.com/news/2016/01/boko-haram-blast-kills-scores-nigeria-maiduguri-160131140615844.html

[43]. K. Uhrmacher and M. Sheridan, "The Brutal Toll of Boko Haram's Attacks on Civilians," The Washington Post, April 3, 2016. https://www.washingtonpost.com/graphics/world/nigeria-boko-haram/

[44]. K. Sieff, "Boko Haram is Forcing More Children to Carry Out Suicide Bombings," The Washington Post, April 12, 2016. https://www.washingtonpost.com/world/africa/un-report-huge-surge-in-boko-haram-child-suicide-bombers-in-west-africa/2016/04/12/b0856ee0-4ac7-49d0-9d3f-9b423631eded_story.html

[45]. "Fleeing Boko Haram, and Dying of Hunger," New York Times 6/24/2016; pg. A5.

[46]. "Genocide in Darfur," United Human Rights Council, May 3, 2016. http://www.unitedhumanrights.org/genocide/genocide-in-sudan.htm

[47]. M. Ray, "Janjaweed Sudanese Militia," http://www.britannica.com/topic/Janjaweed

[48]. "The Lord's Resistance Army," https://www.warchild.org.uk/issues/the-lords-resistance-army

[49]. A. Laing, "Joseph Kony's LRA Abducts Scores of Child Soldiers in New Wave of Attacks," March 3, 2016. http://www.telegraph.co.uk/news/worldnews/joseph-kony/12182422/Joseph-Konys-LRA-abducts-scores-of-child-soldiers-in-new-wave-of-attacks.html

[50]. V. Dobnik, "Wildlife Products May Finance Terrorism," Associated Press, June 16, 2014.

[51]. Ian Niall, <u>The New Poacher's Handbook,</u> The Boydell Press, Suffolk, 1960.

[52]. Collier's Encyclopedia, Macmillan, New York, 1985.

CHAPTER 3.

[1]. V. Braithwaite, <u>Do Fish Feel Pain</u>, Oxford University Press, Oxford, UK, 2010.

[2]. J. Balcombe, <u>What a Fish Knows: The Inner Lives of our Underwater Cousins</u>, Scientific American / Farrar, Strauss and Giroux, New York, 2016.

[3]. Bluefin Tuna Fishing Cruelty Exposed, http://theblackfish.org/news/bluefin-cruelty-exposed/

[4]. A new Mercy For Animals undercover investigation provides a startling glimpse into "Catfish Corner," a fish slaughter facility in Mesquite, Texas http://fish.mercyforanimals.org/

[5]. http://www.nbcdfw.com/news/local/Tougher-Animal-Cruelty-Law-Sought-Over-Fish-Video-114611589.html

[6]. Why is Killing Fish Not Recognized as Animal Cruelty? https://www.reddit.com/r/explainlikeimfive/comments/3fztw5/eli5_why_is_killing_fish_not_recognized_as_animal/

[7]. M. Bekoff, "Fish are Sentient and Emotional Beings and Clearly Feel Pain," Psychology Today, June 19, 2014; https://www.psychologytoday.com/blog/animal-emotions/201406/fish-are-sentient-and-emotional-beings-and-clearly-feel-pain

[8]. "Fish in Tanks," http://www.peta.org/issues/companion-animal-issues/cruel-practices/fish-tanks/

[9]. "Fish Feel Pain, Study Finds," Live Science, April 30, 2009; http://www.livescience.com/7761-fish-feel-pain-study-finds.html

[10]. "Have Animal Rights Groups Finally Lost It? Should Fish Have Rights?" http://www.huffingtonpost.com/mikko-alanne/have-animal-rights-groups_b_811668.html

[11]. C. Brown, "Fish Intelligence, Sentience and Ethics," Animal Cognition 18, 1-17, Jan. (2015).

[12]. "Fish Feel," http://fishfeel.org/

[13]. "Fish Feel Pain; Science Has Repeatedly Shown Us What Common Sense Already Tells Us: Fish Feel Pain and Can Suffer," http://fishfeel.org/resources/facts/

[14]. V. Morell, "Fish Show Signs of Sentience in 'Emotional Fever' Test," SCIENCE Nov. 24, 2015; http://www.sciencemag.org/news/2015/11/fish-show-signs-sentience-emotional-fever-test

[15]. "Cephalopods and Decapod Crustaceans, Their Capacity to Experience Pain and Suffering," Advocates for Animals, 2005.

[16]. C. Brown and K. Laland, "Social Learning in Fishes: a Review," Fish and Fisheries 4, 280-288, (2003).

[17]. C Brown, "Tool Use in Fishes," Fish and Fisheries 13, 105-115, (2012)

[18]. "Ocean Fish Numbers Cut in Half Since 1970," http://www.scientificamerican.com/article/ocean-fish-numbers-cut-in-half-since-1970/

[19]. Coral Reef Alliance, http://coral.org/issue-briefs/

[20]. "Rising Tides, Temperatures, and Costs to Reef Communities," http://coral.org/wordpress/wp-content/uploads/2014/02/climatechange.pdf

[21]. "Undermining the Future Value of Coral Reefs?" http://coral.org/wordpress/wp-content/uploads/2014/02/coralmining.pdf

[22]. "Planning for a Sustainable Reef," http://coral.org/wordpress/wp- content/uploads/2014/02/coastaldev.pdf

[23]. "Short-term Gain, Long-term Loss," http://coral.org/wordpress/wp-content/uploads/2014/02/exploitivefishing.pdf

[24]. The Week, Dec. 12, 2014; pg. 21.

[25]. "Coral Reefs Under Rapid Climate Change and Ocean Acidification," SCIENCE 318, 1737, Dec 14, (2007).

[26]. "One-Third of Reef-Building Corals Face Elevated Extinction Risk from Climate Change and Local Impacts," SCIENCE 321, 560, July 25, (2008).

[27]. "The Great Barrier Reef is Under Attack from El Nino and Climate Change," TIME magazine, April 25, 2016; pg. 14.

[28]. C.A. Burge et al, "Climate Change Influences on Marine Infectious Diseases: Implications for Management and Society," Annu. Rev. Marine Sciences 6, 249-277, (2014).

[29]. Y. Senderovich, I. Izhaki, and M. Halpern (2010), "Fish as Reservoirs and Vectors of Vibrio Cholera." PLoS ONE 5(1): e8607. doi:10.1371/journal.pone.0008607

[30]. L. Novotny et al, "Fish: A Potential Source of Bacterial Pathogens for Human Beings," Vet. Med. 49 (9), 343-358, (2004).

[31]. R.J. Lee et al, "Bacterial Pathogens in Seafood," Food Science, Technology, and Nutrition No. 158, T. Borresen, Editor, Woodhead Publishing Limited, 2008. http://archimer.ifremer.fr/doc/00066/17730/15252.pdf

[32]. "The Silent Suffering of Lobsters," http://www.animalaid.org.uk/h/n/CAMPAIGNS/vegetarianism/ALL/522/

[33]. J. Stewart, "Lobster Diseases," Helgoländer Meeresuntersuchungen 37, 243-254, (1984).

[34]. "Epizootic Shell Disease in Lobsters," https://www.accessscience.com/content/epizootic-shell-disease-in-lobsters/BR0720152

[35]. "Dangers Posed by Eating Undercooked Lobster," https://firstaidcalgary.ca/dangers-posed-by-eating-undercooked-lobster/

[36]. "The Most Dangerous and Lethal Food in the world," http://www.culinary-training.info/blog/the-most-dangerous-and-lethal-food-in-the-world/

CHAPTER 4.

[1]. D. Mozaffarian and EB. Rimm, "Fish Intake, Contaminants, and Human Health: Evaluating the Risks and Benefits," J. Am. Medical Assoc. <u>296</u> (15), 1885, (2006).

[2]. <u>Seafood Choices: Balancing Benefits and Risk</u>, National Academies Press, Committee on Nutrient Relationships in Seafood, Washington D.C., 2007. http://www.nap.edu/read/11762

[3]. https://www.dartmouth.edu/press-releases/risks_benefits_eating_fish_091415.html "Researchers Find Major Gaps in Understanding Risks, Benefits of Eating Fish," Dartmouth University, Sept. 14, 2015.

[4]. http://www.seafoodhealthfacts.org/sites/default/files/final-seafood-health-reference-guide-for-professionals.pdf   Seafood for Health Information for Healthcare Providers A joint project of Oregon State University, Cornell University, and the Universities of California, Delaware, Florida, and Rhode Island.

[5]. L. Fleming et al, "Oceans and Human Health: Emerging Public Health Risks in the Marine Environment," Marine Pollution Bulletin Volume 53, Issues 10-12, pg. 545-560, (2006).

[6]. Causes of Fish Poisoning. MEDIC8, Food Poisoning Guide, http://www.medic8.com/healthguide/food-poisoning/fish-food-poisoning.html

[7]. M. Iwamoto, T. Ayers, B.Mahon and D. Swerdlow, "Epidemiology of Seafood-Associated Infections in the United States," Clinical Microbiology Reviews <u>23</u> (2), 399-411, (2010). http://cmr.asm.org/content/23/2/399.full

[8]. A. Sapkota et al, "Aquaculture Practices and Potential Human Health Risks: Current Knowledge and Future Priorities," Environment International 34 (8), 1215-1226, Nov. (2008).

[9]. "Health Risks of Seafood Fraud," http://usa.oceana.org/health-risks-seafood-fraud

[10]. "A Menace Afloat," New York Times July 19, 2016; pg. D1.

[11]. http://www.greenfacts.org/en/mercury/l-2/mercury-2.htm

[12]. "Health Effects of Exposures to Mercury," https://www.epa.gov/mercury/health-effects-exposures-mercury.

[13]. A.Marques et al, "Climate Change and Seafood Safety: Human Health Implications," Food Research International 43, 1766-1779, (2010). http://climateknowledge.org/Food_Water_Illness_Models/Marques_Climate_Change_Seafood_FoodResInt_2010.pdf

[14]. "Human Health Risks," http://www.centerforfoodsafety.org/issues/312/aquaculture/human-health-risks#

[15]. "Dioxins and their Effects on Human Health," World Health Organization, http://www.who.int/mediacentre/factsheets/fs225/en/

[16]. "PCBs Are a Probable Human Carcinogen," http://www.clearwater.org/news/pcbhealth.html

[17]. S. Etheridge, "Paralytic Shellfish Poisoning: Seafood Safety and Human Health Perspectives," Toxicon 56 (2), 108, (2010).

[18]. M.M. Storelli, "Potential Human Health Risks from Metals (Hg, Cd, and Pb) and Polychlorinated Biphenyls (PCBs) via Seafood Consumption: Estimation of Target Hazard Quotients

(THQs) and Toxic Equivalents (TEQs)," Food and Chemical Toxicology 46 (8), 2782-2788, (2008).

[19]. G. Amagliani et al, "Incidence and Role of Salmonella in Seafood Safety," Food Research Internat. 45 (2), 780-788, (2012).

[20]. R. J. Wittman and G.J. Flick, "Microbial Contamination of Shellfish: Prevalence, Risk to Human Health, and Control Strategies," Annual Rev. of Public Health 16, 123-140, (1995).

[21]. F. Feldhusen, "The Role of Seafood in Bacterial Foodborne Diseases," Microbes and Infection 2 (13), 1651-1660, (2000).

[22]. M. Greger, MD, How Not To Die, Flatiron Books, New York, 2015.

[23]. M. Greger, MD, "Fish," nutritionfacts.org/topics/fish

[24]. G. Kimbrell, S. Wu, and C. Stella, "Industrial Aquaculture: It's Déjà Vu All Over Again, but this Time We Can Choose a Different Path," Feb. 25, 2016. www.centerforfoodsafety.org/blog/4260/industrial-aquaculture-its-dj-vu-all-over-again-but-this-time-we-can-choose-a-different-path#

[25]. "Threats to the Environment and Wildlife," Center for Food Safety, www.centerforfoodsafety.org/issues/312/aquaculture/threats-to-environment-and-wildlife

[26]. C. Harvel et al, "Emerging Marine Diseases—Climate Links and Anthropogenic Factors," SCIENCE 285, 1505, Sept 3, (1999). http://digitalcommons.unl.edu/cgi/viewcontent.cgi?article=1451&context=parasitologyfacpubs

[27]. C. Burge ,et al, "Climate Change Influences on Marine Infectious Diseases: Implications for Management and Society," Annual Rev. Marine Science 6, 249-277 (2014). http://www.

annualreviews.org/eprint/qtrxqRIr8GERt4WyRUDC/full/10.1146/
annurev-marine-010213-135029

CHAPTER 5.

[1]. N. Oreskes and E. Conway, <u>Merchants of Doubt</u>, Bloomsbury Press, New York, 2010.

[2]. "Global Warming: Faster Than Expected," Scientific American, November 2012.

[3]. Global Warming Effects Map, http://environment.nationalgeo-graphic.com/environment/global-warming/gw-impacts-interactive/

[4]. "Earth Temperature in 1998 Is Reported at Record High," New York Times, 12/18/1998.

[5]. "Earth's Temperature Shot Skyward in 1998," Science News **155**, 6, Jan 2, (1999).

[6]. "Global Temperature at a High For the First 5 Months of 1998," New York Times 6/8/1998.

[7]. "Warming's Unpleasant Surprise: Shivering in the Greenhouse," SCIENCE <u>281</u>, 156, July 10, (1998).

[8]. "The Mercury's Rising," Newsweek December 4, 2000; pg. 52.

[9]. "Global Temperatures Highest in 4000 Years," New York Times 3/8/2013; pg. A15.

[10]. "2015 Far Eclipsed 2014 As World's Hottest Year, Climate Scientists Say," New York Times 1/21/2016; pg. A1

[11]. "For Third Year, The Earth in 2016 Set Heat Record," New York Times 1/29/2017; pg. A1.

[12]. "Unrelenting California Heat Wave Is Blamed for More Than 100 Deaths," New York Times 7/28/2006, pg. A24.

[13]. "Iran Heat Index Near Record 163 Degrees, Fatalities Reported," http://www.digitaljournal.com/news/world/iran-heat-index-near-record-165-degrees-fatalities-reported/article/439879

[14]. "Hottest. Year. Ever," The Week, 8/5/2016, pg. 19.

[15]. Hugh Naylor, "An Epic Middle East Heat Wave Could Be Global Warming's Hellish Curtain-raiser," The Washington Post, August 10, 2016.

[16]. "The Impacts of Climate Change On Human Health in the United States, A Scientific Assessment," U.S. Global Change Research Program, April, 2016. health2016.globalchange.gov

[17]. "Deadly Heat is Forecast in Persian Gulf by 2100," New York Times 10/27/2015; pg. A6.

[18]. "A Surge in Heat Wave "Danger Days" Is Expected in Coming Decades," http://www.scientificamerican.com/article/a-surge-in-heat-wave-danger-days-is-expected-in-coming-decades-infographic/

[19]. "Spiking Temperatures in the Arctic Startle Scientists; A Vicious Cycle as Sea Ice is Lost," New York Times 12/22/2016; pg. A4.

[20]. "Progress Seen On Warming, With a Caveat," New York Times 9/28/2015; pg. A1.

[21]. "If We Dig Out All Our Fossil Fuels, Here's How Hot We Can Expect It to Get," New York Times 4/8/2015. http://nyti.ms/1a5hioj

[22]. "Effect of Heat Waves on Animals," Biometeorology, 7/24/2016; http://www.climate-policy-watcher.org/biometeorology/effect-of-heat-waves-on-animals.html

[23]. "Animal Winners and Losers of Summer's Heat Waves," http://news.nationalgeographic.com/news/2012/09/pictures/120910-animals-heat-wave-drought-global-warming-science-enviro/

[24]. "Weather Conditions and Nonhuman Animals," http://www.animal-ethics.org/weather-conditions-nonhuman-animals/

[25]. https://www.cbsnews.com/news/carbon-dioxide-emissions-rise-to-24-million-pounds-per-second

[26]. "Deforestation and Greenhouse-Gas Emissions," www.cfr.org/forests-and-land-management/deforestation-greenhouse-gas-emissions/p14919

[27]. https://www.nifc.gov/safety/safety_documents/Fatalities-by-Year.pdf

[28]. http://wildfiretoday.com/2016/10/11/study-concludes-climate-change-has-doubled-acres-burned-in-western-u-s/

[29]. J. Abatzoglou and A. Williams, "Impact of Anthropogenic Climate Change on Wildfire Across Western U.S. Forests," Proc. Nat. Acad. of Sciences, www.pnas.org/content/113/42/11770

[30]. "Half of Rise in Fire Risk is Tied to Climate Change," New York Times 10/11/2016; pg. A11.

[31]. "Fire Season? In Some Spots, It's Year-Round," New York Times 4/13/2016; pg. A1.

[32]. "After The Burn; Drought-Fueled Wildfires May Be Damaging Some Forests Beyond Recovery," New York Times, (Science Times) 9/22/2015.

[33]. http://www.iii.org/fact-statistic/wildfires

[34]. "What's the true cost of the American West's wildfires?" http://fortune.com/2015/09/15/cost-california-wildfires/

[35]. The Week, October 3, 2014.

[36]. "Rising Sea Levels More Dangerous Than Thought," Scientific American 8/27/2015. https://www.scientificamerican. com/article/rising-sea-levels-more-dangerous-than-thought/

[37]. "Arctic Sea Ice Minimum," http://climate.nasa.gov/ vital-signs/arctic-sea-ice/

[38]. "Melting of Arctic Sea Ice Already Setting Records in 2016," EcoWatch, http://www.ecowatch.com/melting-of-arctic-sea-ice-already-setting-records-in-2016-1891079333.html

[39]. "Open Water at Pole Is Not Surprising, Experts Say," New York Times 8/29/2000.

[40]. "Dead Trees and Shriveling Glaciers as Alaska Melts," New York Times 8/18/1998.

[41]. "The Race to Alaska Before it Melts," New York Times 6/26/2005; Travel, pg.1.

[42]. "As Alaska's Glaciers Melt, Landslides and Tsunamis Follow," New York Times 7/6/2016, pg. A10.

[43]. "Melting Mountain Majesties: Warming in Austrian Alps," New York Times 8/8/2005.

[44]. "Global Warming Poses Threat to Ski Resorts in the Alps," New York Times 12/16/2006.

[45]. "As Andean Glaciers Shrink, Water Worries Grow," New York Times 11/24/2002.

[46]. "Retreat of Huge Glacier in China Signals Trouble for Asian Water Supply," New York Times 12/9/2015; pg. A4.

[47]. "On Thinning Ice, Are The World's Glaciers in Mortal Danger?" Science News 164, 215 (2003); (www.sciencenews.org)

[48]. B. Dennis and C. Mooney, "Scientists Nearly Double Sea Level Rise Projections for 2100, Because of Antarctica," The Washington Post, March 30, 2016; https://www.washingtonpost. com/news/energy-environment/wp/2016/03/30/antarctic-loss-could-double-expected-sea-level-rise-by-2100-scientists-say/?utm_term=.5e21a5bbeae3

[49]. "Tourists at the End of the World," The Nation, 11/9/2015, pg. 12.

[50]. "Satellite-Observed Changes in the Arctic," Physics Today, August 2004, pg. 38.

[51]. N. Jones, "Abrupt Sea Level Rise Looms As Increasingly Realistic Threat," May 5, 2016; http://e360.yale.edu/feature/abrupt_sea_level_rise_realistic_greenland_antarctica/2990/

[52]. "What is the Relationship Between Glaciers and Sea Level?" U.S. Geological Survey, https://www2.usgs.gov/climate_landuse/glaciers/glaciers_sea_level.asp
https://www2.usgs.gov/faq/categories/9750/3476

[53]. "An Icy Riddle As Big As Greenland," New York Times 6/8/2004, pg. F1.

[54]. "In Ancient Ice Ages, Clues to Climate," New York Times 2/16/1999; pg. F1.

[55]. "Using Up Earth's Fossil Fuels Would Destroy All Ice, Research Says," New York Times 9/12/2015; pg. A10

[56]. "Melting of Great Ice Sheet May Threaten Coastal Cities by 2100," New York Times 3/31/2016; pg. A1.

[57]. "Climate Change Countdown," New York Times 4/5/2016; pg. D1.

[58]. "Worsened by Climate Change, 'King Tides' Transform Florida Life," New York Times 11/18/2016; pg. A1.

[59]. "In Pacific, Growing Fear of Paradise Engulfed," New York Times 3/2/1997; pg. A1.

[60]. "Remote Pacific Nation, Threatened by Rising Seas," New York Times 7/3/2016; pg. 10.

[61]. "Rising Seas to Endanger Cities," The Week, Sept. 25, 2015; pg. 19.

[62]. C. Katz, "Melting Glaciers Liberate Ancient Microbes, The Release of Life-forms in Cold Storage for Eons Raises New Concerns about the Impacts of Climate Change," https://www.scientificamerican.com/article/melting-glaciers-liberate-ancient-microbes/

[63]. "Even in Frigid North, Hints of Warmer Temperatures," New York Times 10/10/2000; pg. F1.

[64]. "Big Artic Perils Seen in Warming," New York Times 10/30/2004, pg. A1.

[65]. "The Climate Refugees of the Arctic," New York Times 12/20/2016, pg. A26.

[66]. "Climate Change Impacts in the United States," nca2014.globalchange.gov

[67]. "The Freak Weather," E Magazine, March/April 2012.

[68]. "Studies Look for Signs of Climate Change in 2014's Extreme Weather Events," New York Times 11/6/2015; pg. A16.

[69]. "Taking Global Warming to the People," SCIENCE 283, 1440 (1999).

[70]. "Scientists Say Earth's Warming Could Set Off Wide Disruptions," New York Times 9/18/1995; pg. A1.

[71]. "In China, Encroaching Seas of Sand," New York Times 12/24/2016; pg. A1.

[72]. "Drought Cuts Short an African Success Story," New York Times 4/13/2016; pg. A1.

[73]. "California Drought Is Made Worse by Global Warming, Scientists Say," New York Times 4/1/2015; New York Times 8/20/2015.

[74]. "A Wet Winter Won't Save California," New York Times 9/19/2015.

[75]. "Climate Scientists Forecast More Floods Like Louisiana's," New York Times 9/8/2016; pg. A19.

[76], "The Coming Megafloods," Scientific American, January, 2013, pg. 64.

[77]. "Global Warming: The Culprit?" TIME Magazine 10/3/2005; pg. 43.

[78]. "Global Warming is Expected to Raise Hurricane Intensity," New York Times 9/30/2004. "Hurricanes and Climate Change," https://www.c2es.org/science-impacts/extreme-weather/hurricanes

[79]. W. Michener et al, "Climate Change in Hurricanes and Tropical Storms, and Rising Sea Level in Coastal Wetlands," Ecological Applications, 7(3), 770-801 (1997).

[80]. "Extreme Weather: The New Normal," The Week, 11/18/2016, pg. 14.

[81]. "Climate Change: Why Poor Nations Suffer Most," The Week 11/29/2013; pg. 15.

[82]. P. Smith and P.J. Gregory, "Climate Change and Sustainable Food Production," Proc. Nutrition Society 72, 21-28, (2013).

[83]. "Livestock's Long Shadow," Food and Agricultural Organization of the United States, Rome, 2006.

[84]. W.F. Laurance, "A Crisis in the Making: Responses of Amazon Forests to Land Use and Climate Change," TREE 13 (10). 411-415, (1998).

[85]. R. Adams, "Global Climate Change and Agriculture: An Economic Perspective," Am. J. Agric. Econ. 71 (5), 1272-1279, (1989).

[86]. "Climate Change and Agriculture, A Review of Impacts and Adaptations," The World Bank Environment Department, June, 2003.

[87]. "Climate Impacts on Agriculture and Food Supply," U.S. Environmental Protection Agency, https://www.epa.gov/climate-impacts/climate-impacts-agriculture-and-food-supply

[88]. "Global Climate Change Impacts in the United States," U.S. Global Change Research Program, 2009, https://nca2009.global-change.gov/index.html

[89]. "Impacts of Climate Change on Human Health in the United States, Food Safety, Nutrition, and Distribution," U.S. Global Change Research Program, 2016. https://health2016.globalchange.gov

[90]. "Climate Change, Evidence, Impacts, and Choices," National Research Council, National Academy of Sciences, 2012. Climate Change-2012 NRC report.pdf

[91]. "Panel's Warning on Climate Risk: Worst is Yet to Come," New York Times 3/31/2014, pg. A1

[92]. "Climate Change Seen Posing Risk to Food Supplies," New York Times 11/2/2013, pg. A1 (quoting from IPCC 2013 report).

[93]. R. Goodland and J. Anhang, "Livestock and Climate Change," World Watch, Nov/Dec 2009; www.worldwatch.org

[94]. "Growing Greenhouse Emissions Due To Meat Production," UNEP Global Environment Alert Service (GEAS), www.unep.org/geas

[95]. "The World's Leading Driver of Climate Change: Animal Agriculture," new harvest, January 18, 2015; http://www.new-harvest.org/the_world_s_leading_driver_of_climate_change_animal_agriculture

[96]. "G. Koneswaran and D. Nierenberg, "Global Farm Animal Production and Global Warming: Impacting and Mitigating Climate Change," Environ. Health Perspect. 116 (5), 578-582 (2008).

[97]. C. Cederberg et al, "Trends in Greenhouse Gas Emissions From Consumption and Production of Animal Food Products – Implications For Long-Term Climate Targets," Animal 7 (2), 330-340 (2013). doi:10.1017/S1751731112001498

[98]. A. McMichael et al, "Food, Livestock Production, Energy, Climate Change, and Health," Lancet 370, 1253-1263 (Oct. 6, 2007)

[99]. "Ban Ki-moon Calls on New Generation to Take Better Care of Planet Earth Than His Own," http://www.un.org/apps/news/story.asp?NewsID=21720&Cr=global&Cr1=warming#.WKxTUlMrKpo

[100]. National Wildlife Federation, "Global Warming and Wildfires," http://www.nwf.org/Wildlife/Threats-to-Wildlife/Global-Warming/Global-Warming-is-Causing-Extreme-Weather/Wildfires.aspx

[101]. "Negative Effects of Wildfire on Wildlife," https://www.thewildlifemuseum.org/exhibits/burning-for-wildlife/negative-effects-of-wildfire-on-wildlife/

[102]. S. Zielinski, "What Do Animals Do in a Wildfire?" National Geographic, 7/22/2014; http://news.nationalgeographic.com/news/2014/07/140721-animals-wildlife-wildfires-nation-forests-science/

[103]. H. Janssen, "5 Unexpected Ways That Wildfires Affect Wildlife, Ecosystems of Forests," AccuWeather, 6/10/2016; http://

www.accuweather.com/en/weather-news/five-effects-wildfire-west-forests-wildlife-ecology-animals-plants/53289791

[104]. D. Inkley, "5 Ways Wildfires Threaten Western Wildlife," June 19, 2012; http://blog.nwf.org/2012/06/5-ways-wildfires-threaten-western-wildlife/

[105]. L. Zuckerman, "Massive Wildfires in U.S. Northwest Destroyed Habitats, Threaten Wildlife," https://www.scientificamerican.com/article/massive-wildfires-in-u-s-northwest-destroyed-habitats-threaten-wildlife/

[106]. M. Denchak, "Are the Effects of Global Warming Really that Bad?" National Resources Defense Council, 3/15/2016; https://www.nrdc.org/stories/are-effects-global-warming-really-bad

[107]. "Animals in Natural Disasters," http://www.animal-ethics.org/animals-natural-disasters/

[108]. J. John, "Hotter Days Could Kill More Desert Songbirds," The Wildlife Society, 3/1/2017; http://wildlife.org/hotter-days-could-kill-more-desert-songbirds/

[109]. J. Hayes, "Heatwaves to Hit Wildlife Hard," ABC Environment 3/13/2013; www.abc.net.au/environment/articles/2013/03/12/3711481.htm

[110]. L. O'Connor, "Get Used To These Extreme Summer Heat Waves," Huffington Post, 07/24/2016; http://www.huffingtonpost.com/entry/summer-heat-wave-climate-change_us_57951438e4b02d5d5ed1f7eb

[111]. "Wildlife, Weather, and Heat," http://www.creativecowboyfilms.com/blog_posts/wildlife-weather-and-heat

[112]. "100,000 Bats Fall Dead From the Sky During a Heatwave in Australia," http://www.telegraph.co.uk/news/worldnews/austra-liaandthepacific/australia/10558183/100000-bats-fall-dead-from-the-sky-during-a-heatwave-in-Australia.html

[113]. D. Rose, "Wildlife, Climate Change and Mass Death," January, 2014; http://extinctionstudies.org/2014/01/16/wildlife-climate-change-and-mass-death/

[114]. N. Thompson, "Killer Heatwave Wreaks Havoc in Southeast Asia," CNN, May 12, 2016; http://www.cnn.com/2016/05/12/homepage2/southeast-asia-drought-el-nino/

[115]. M. Roosevelt, "State Prepares to Deal With Heat Waves, Flooding, Wildlife Die-offs and Other Expected Results of Climate Change," LA Times, 8/4/2009; http://articles.latimes.com/2009/aug/04/local/me-climate4

[116]. "Does Extreme Heat Affect Farm Animals?" http://kb.rspca.org.au/does-extreme-heat-affect-farm-animals_562.html

[117]. "Effect of Heat Waves on Animals," Biometeorology, 7/24/2016; http://www.climate-policy-watcher.org/biometeorology/effect-of-heat-waves-on-animals.html

[118]. "2006 North American Heat Wave," https://en.wikipedia.org/wiki/2006_North_American_heat_wave

[119]. A. Carr and S. Breslin, "330 Million Impacted By India Heat Wave That Has Killed At Least 160, Officials Say," 4/21/2016; https://weather.com/safety/heat/news/deadly-southern-eastern-india-heat-wave-2016

[120]. R. Inani, "More Than 2,300 People Have Now Died in India's Heat Wave," TIME

World, 6/1/2015; http://time.com/3904590/
india-heatwave-monsoon-delayed-weather-climate-change/

[121]. "Climate Change - Effects on Animals, Birdlife and Plants,"
http://www.climateandweather.net/global-warming/climate-
change-and-animals.html

[122]. C. Welch, "What Animals Are Likely to Go Extinct First
Due to Climate Change," National Geographic, 4/30/2015.

[123]. T.L. Root and S.H. Schneider, <u>Climate Change: Overview
and Implications for Wildlife,</u> Island Press, Washington DC, 2002.

[124]. "Hurricanes in History," National Hurricane Center, http://
www.nhc.noaa.gov/outreach/history/

[125]. "11 Facts About Hurricane Katrina," https://www.dosome-
thing.org/us/facts/11-facts-about-hurricane-katrina

[126]. "Hurricane Matthew Death Toll Rises. 'Difficult Days'
Ahead, N.C. Governor Says," NBC News, 10/15/2016; http://
www.nbcnews.com/storyline/hurricane-matthew/hurricane-mat-
thew-death-toll-rises-difficult-days-ahead-n-c-n666891

[127]. P. Wright, "Haitians Struggle to Survive in Wake of
Hurricane Matthew," The Weather Channel, 10/13/2016; https://
weather.com/news/news/hurricane-matthew-haiti-latest-news-0

[128]. P. Wright, "Haiti Prepares For a 'Tsunami of Cholera' After
Hurricane Matthew," The Weather Channel, 10/4/2016; https://
weather.com/health/news/haiti-hurricane-matthew-cholera

[129]. T. Philpott, "Hurricane Matthew Killed Millions of Farm
Animals in North Carolina. (It also likely caused massive amounts
of toxic hog poop to flow into rivers and streams)," Mother Jones,

10/14/2016; http://www.motherjones.com/environment/2016/10/
hurricane-matthew-killed-animals-hog-poop

[130]. P. Archer, "Countless Animals Killed,
Injured in Fort Bend County Flood Waters,"
6/9/2016; http://www.click2houston.com/news/
countless-animals-killed-injured-in-fort-bend-county-flood-waters

[131]. E.O. Wilson, Half Earth, Our Planet's Fight for Life,
Liveright Publishing Corp., New York, 2016.

[132]. S. Zielinski, "Climate Change Will Accelerate Earth's Sixth
Mass Extinction," Smithsonian, April 30, 2015; http://www.smith-
sonianmag.com/science-nature/climate-change-will-accelerate-
earths-sixth-mass-extinction-180955138/#uLx8FC7e3ZB5fcoF.99

[133]. S. Zielinski, "One in Five Lizard Species Could
Be Extinct by 2080," Smithsonian, May 14, 2010;
http://www.smithsonianmag.com/science-nature/
one-in-five-lizard-species-could-be-extinct-by-2080-28461208/

[134]. "The Extinction Crisis," Center For Biological Diversity,
http://www.biologicaldiversity.org/programs/biodiversity/
elements_of_biodiversity/extinction_crisis/

[135]. "Extinction Risk From Global Warming," https://
en.wikipedia.org/wiki/Extinction_risk_from_global_warming

[136]. C. Thomas et al, "Extinction Risk From Climate Change,"
Nature 427, 145-148, Jan 8, (2004).

[137]. C. Welch, "What Animals Are Likely to Go Extinct First
Due to Climate Change," National Geographic, 4/30/2015.

[138]. M. Urban, "Accelerating Extinction Risk From Climate
Change," SCIENCE 348, 571-573, May1, (2015).

[139]. "Climate Change Impacts, Wildlife at Risk; One-fourth of the Earth's Species Could be Headed for Extinction by 2050 Due to Climate Change," https://j02alinsangan. wordpress.com/2016/02/17/climate-change-threats-and-impact/ https://www.nature.org/ourinitiatives/urgentissues/ global-warming-climate-change/threats-solutions/

[140]. "Climate Change - Effects on Animals, Birdlife and Plants," http://www.climateandweather.net/global-warming/climate-change-and-animals.html

[141]. J. Lawler et al, "Projected Climate-Induced Faunal Change in the Western Hemisphere," Ecology 90 (3), 588-597 (2009).

[142]. C. Dell'Amore, "7 Species Hit Hard by Climate Change—Including One That's Already Extinct; Coral, Polar bears, and Frogs are Among the Species Hit Hardest," National Geographic April 2, 2014.

[143]. "Climate Impacts on Ecosystems: Overview," United States Environmental Protection Agency, https://www.epa.gov/ climate-impacts/climate-impacts-ecosystems

[144]. "Ecological and Evolutionary Responses to Recent Climate Change," Annu. Rev. Ecol. Evol. Syst. 37, 637-669 (2006).

[145]. "Wildlife in a Warming World," http://www.nwf.org/

[146]. M. Townsend, "5 Animals at Risk of Extinction Because of Climate Change (Besides Polar Bears)", http://www.care2.com/ causes/5-animals-at-risk-of-extinction-because-of-climate-change-besides-polar-bears.html#ixz3mBsgdsxE

[147]. "Hawaii on Verge of Worst Coral Bleaching in History as Water Temperatures Soar," EcoWatch, Sept. 16, 2005; http://www.

care2.com/causes/hawaii-on-verge-of-worst-coral-bleaching-in-history-as-water-temperatures-soar.html#ixzz3mBtqK2Qs

[148]. M. Innis, "Great Barrier Reef Facing Multiple Threats, Report Says," New York Times 12/2/2016, pg. A6.

[149]. "The Consequences of Global Warming On Wildlife," Natural Resources Defense Council, http://www.nrdc.org/global-warming/fcons/fcons3.asp

[150]. "Shed a Tear for the Reefs," New York Times 3/19/2017; pg. SR10.

[151]. D. Cave and J. Gillis, "Great Barrier Reef Is Imperiled, Much Of It Dying or Dead," New York Times 3/16/2017; pg. A8.

[152] "Malaria: Affects Animals as Well as Humans," http://www.abc.net.au/radionational/programs/scienceshow/malaria-affects-animals-as-well-as-humans/3130296.

[153]. "One in Four U.S. Deer Is Infected With Malaria," Smithsonian.com, http://www.smithsonianmag.com/smithsonian-institution/one-in-four-deer-infected-malaria-180958046/

[154]. J. Viegas, "Ebola's Deadly Jump From Animal to Animal," July 30, 2014; http://www.seeker.com/ebolas-deadly-jump-from-animal-to-animal-1768901643.html

[155]. "How Are Animals Involved in Ebola Outbreaks?" CDC, Centers for Disease Control and Prevention, Feb. 18, 2016; https://www.cdc.gov/vhf/ebola/transmission/qas-pets.html

[156]. D. Biello, "Deadly by the Dozen: 12 Diseases Climate Change May Worsen; The Wildlife Conservation Society has identified some of the illnesses that global warming may

exacerbate," https://www.scientificamerican.com/article/twelve-diseases-climate-change-may-make-worse/

[157]. M. Walker, "Gorillas and Chimps are Threatened by Human Disease," BBC Nature, http://www.bbc.co.uk/nature/19408683

[158]. C. Curtin, "Ebola Epidemic Wiping Out Gorilla Populations," Dec. 7, 2006; https://www.scientificamerican.com/article/ebola-epidemic-wiping-out/

[159]. F. Schaumburg et al, "Drug-Resistant Human Staphylococcus Aureus in Sanctuary Apes Pose a Threat to Endangered Wild Ape Populations," American Journal of Primatology 74, 1071-1075, (2012).

[160]. C. Nunni and B. Hare, "Pathogen Flow: What We Need to Know," American Journal of Primatology 74, 1084-1087, (2012).

[161]. "Simian Immunodeficiency Virus," https://en.wikipedia.org/wiki/Simian_immunodeficiency_virus

[162]. S. Lovgren, "HIV Originated With Monkeys, Not Chimps, Study Finds," National Geographic News June 12, 2003; http://news.nationalgeographic.com/news/2003/06/0612_030612_hivvirusjump.html

[163]. "West Nile Virus, Fact Sheet," World Health Organization, July, 2011; http://www.who.int/mediacentre/factsheets/fs354/en/

[164]. "West Nile Encephalitis," Center for Food Security and Public Health, www.cfsph.iastate.edu/FastFacts/pdfs/West_nile_fever_F.pdf

[165]. L. Marsa, "The Zone; Dengue in Texas, Malaria in New York, Hypertoxic Pollen in Baltimore. Climate Change is

Scrambling Disease in Scary New Ways," DISCOVER magazine, December, 2010; pg. 39ff.

[166]. K. Rogers, "On Top of Everything Else, Climate Change Helps West Nile to Spread," Feb. 7, 2017. https://motherboard. vice.com/en_us/article/3d73ew/on-top-of-everything-else-climate-change-helps-west-nile-disease-to-spread

[167]. J. Bryner, "Animal-to-Human Diseases Kill 2.2 Million People Each Year," Live Science, July 6, 2012; http://www.li-vescience.com/21426-global-zoonoses-diseases-hotspots.html

[168]. "Dengue Virus Circulating Between Monkeys and Mosquitoes Could Emerge to Cause Human Outbreaks," https://www.sciencedaily.com/releases/2011/06/110613142235.htm

[169]. "Cholera; Fast Facts," www.cfsph.iastate.edu/FastFacts/pdfs/cholera_F.pdf

[170]. W. Stevens, "Warmer, Wetter, Sicker: Linking Climate to Health," New York Times 8/10/1998; pg. A1.

[171]. A.J. McMichael et al, "Climate Change and Human Health: Risks and Responses," World Health Organization, Geneva, 2003. http://www.who.int/globalchange/publications/cchhbook/en/

[172]. "Plague," National Geographic, http://www.nationalgeo-graphic.com/science/health-and-human-body/human-diseases/the-plague/

[173]. "Plague, Fast Facts," Center for Food Safety and Public Health, www.cfsph.iastate.edu/FastFacts/pdfs/Plague_F.pdf

[174]. T. Nguyen, "Famous People You Didn't Know Had TB," Nov. 2, 2015; http://www.voxmagazine.com/news/science/

famous-people-you-didn-t-know-had-tb/article_b57905f2-beff-570d-a9a0-e335df9e6300.html

[175]. "Bovine Tuberculosis," Center for Food Safety and Public Health, www.cfsph.iastate.edu/FastFacts/pdfs/bovine-tuberculosis_F.pdf

[176]. "Global Warming and Human Health," Climate Institute, www.climate.org/topics/health/html

[177]. "Climate Change Impacts; Climate Impacts on Human Health," U.S. Environmental Protection Agency, https://www.epa.gov/climate-impacts/climate-impacts-human-health

[178]. "The Impacts of Climate Change on Human Health in the United States, A Scientific Assessment," U.S. Global Change Research Program, 2016, Executive Summary and Full Report, health2016.globalchange.gov

[179]. "Climate Warming and Disease Risks for Terrestrial and Marine Biota," SCIENCE 296, 2158, June 21, (2002).

[180]. The Week, July 17, 2015; pg. 19.

[181]. D. Harvell et al, "Climate Change and Wildlife Diseases: When Does the Host Matter the Most?" Ecology 90 (40), 912-920, (2009).

[182]. M. Gallana et al, "Climate Change and Infectious Diseases of Wildlife: Altered Interactions Between Pathogens, Vectors, and Hosts," Current Zoology 59 (3), 427-437, (2013).

[183]. A. Fiegan Durbin, "Chikungunya Virus on the Move," Harvard University, Special Edition on Infectious Diseases, Dec. 31, 2014; http://sitn.hms.

harvard.edu/flash/special-edition-on-infectious-disease/2014/
chikungunya-virus-on-the-move/

[184]. "Climate Change Impacts: Climate Impacts on
Agriculture and Food Supply," U.S. Environmental
Protection Agency, https://www.epa.gov/climate-impacts/
climate-impacts-agriculture-and-food-supply

[185]. "As Oceans Warm, Problems From Viruses and Bacteria
Mount," New York Times 1/24/1999; pg. 15.

[186]. "Climate Change and Wildlife Health: Direct and Indirect
Effects," U.S. Geological Survey, https://www.nwhc.usgs.gov/pub-
lications/fact_sheets/pdfs/Climate_Change_and_Wildlife_Health.
pdf

[187]. K. Acevedo-Whitehouse and A. Duffus, "Effects of
Environmental Change on Wildlife Health," Phil. Trans. R. Soc. B
364, 3429-3438, (2009).

[188]. R. Sinha et al, "Meat Intake and Mortality," Arch. Intern.
Med. 169 (6), 562-571, (2009).

[189]. T. Huang et al, "Cardiovascular Disease Mortality and
Cancer Incidence in Vegetarians: A Meta-Analysis and Systematic
Review," Ann. Nutr. Metab. 60, 233-240, (2012).

[190]. M. Thorogood et al, "Risk of Death From Cancer and
Ischaemic Heart Disease in Meat and Non-meat Eaters," BMJ
(British Medical Journal) 308, 1667, (1994).

[191]. J. Wise, "Vegetarians Have Lower Risk of Colorectal
Cancers, Study Finds," BMJ 350, h1313, (2015); http://www.
bmj.com/content/350/bmj.h1313?utm_source=TrendMD&utm_
medium=cpc&utm_campaign=TBMJ_UK_TrendMD-0

[192]. D. Rakel, "Vegetarian Dietary Patterns and the Risk of Colorectal Cancers," JAMA internal medicine; http://www.practiceupdate.com/content/vegetarian-dietary-patterns-and-the-risk-of-colorectal-cancers/22833?trendmd-shared=1

[193]. "The War on Delicious," TIME magazine, Nov. 9, 2015; pg. 30.

[194]. R. Hubbard et al, "The Potential of Diet to Alter Disease Processes," Nutrition Research 14 (12), 1853-1896, (1994).

[195]. T. Philpott, "Playing Chicken," Mother Jones, May / June 2016, pg. 41.

[196]. D. McNeil Jr., "Deadly Superbugs Pose a Huge Threat, W.H.O. Says," New York Times 2/28/2017; pg. A8.

[197]. "The Antibiotics Crisis," The Week, Nov. 22, 2013; pg. 9.

[198]. "Global Greenhouse Gas Emissions Data," U.S. Environmental Protection Agency, https://www.epa.gov/ghgemissions/global-greenhouse-gas-emissions-data

[199]. T. Philpott, "Burger With Flies," Mother Jones, March/April 2017; pg. 64.

[200]. "Climate Change: How Do We Know?" https://climate.nasa.gov/evidence/

[201]. "Assessing the Global Climate in March 2017," National Oceanic and Atmospheric Administration, National Centers for Environmental Information, https://www.ncei.noaa.gov/news/global-climate-201703

[202]. "Climate Change on Pace to Occur 10 Times Faster Than Any Change Recorded in Past 65 Million Years, Stanford Scientists

Say," Stanford Report, August 1, 2013; http://news.stanford.edu/news/2013/august/climate-change-speed-080113.html

[203]. "How is Today's Warming Different from the Past?" https://earthobservatory.nasa.gov/Features/GlobalWarming/page3.php

[204]. N. Diffenbaugh and C. Field, "Changes in Ecologically Critical Terrestrial Climate Conditions," SCIENCE 341 (Issue 6145), 486-492, Aug 2, (2013).

[205]. A. Mulkern, "Today's Climate Change Proves Much Faster Than Changes in Past 65 Million Years," Climate Wire, Scientific American, Aug. 2, 2013; https://www.scientificamerican.com/article/todays-climate-change-proves-much-faster-than-changes-in-past-65-million-years/

[206]. "Climate Change Science. Causes of Climate Change," U.S. Environmental Protection Agency, https://www.epa.gov/climate-change-science/causes-climate-change

[207]. O. Hoegh-Guldberg et al, "Coral Reefs Under Rapid Climate Change and Ocean Acidification," SCIENCE 318 (Issue 5857), 1737-1742, Dec 14, (2007). www.sciencemag.org

[208]. D. Spittlehouse and R. Stewart, "Adaptation to Climate Change in Forest Management," BC Journal of Ecosystems and Management 4 (1), 1-7, (2003).

[209]. C. Burge et al, "Climate Change Influences on Marine Infectious Diseases: Implications for Management and Society," Annual Review Marine Science 6, 249-277, (2014).

CHAPTER 6.

[1]. "Nuclear Arsenals," International Campaign to Abolish Nuclear Weapons, http://www.icanw.org/the-facts/nuclear-arsenals/#

[2]. O. Toon, A. Robock, and R. Turco, "Environmental Consequences of Nuclear War," Physics Today, Dec. 2008, pg. 37-42. http://climate.envsci.rutgers.edu/pdf/ToonRobockTurcoPhysicsToday.pdf

[3]. S. Starr, "Catastrophic Climatic Consequences of Nuclear Conflict," International Network of Engineers And Scientists Against Proliferation, INESAP Bulletin, 2009; http://www.nucleardarkness.org/warconsequences/catastrophicclimaticconsequences/

[4]. S. Starr, "Nuclear War: An Unrecognized Mass Extinction Event Waiting To Happen," New York Academy of Medicine, Feb 28, 2015; https://ratical.org/radiation/NuclearExtinction/StevenStarr022815.html

[5]. A. Robock and O. Brian Toon, "Local Nuclear War," http://climate.envsci.rutgers.edu/pdf/RobockToonSciAmJan2010.pdf

[6]. "How Nuclear Bombs Affect the Environment," http://education.seattlepi.com/nuclear-bombs-affect-environment-6173.html

[7]. T. Mousseau, "At Chernobyl and Fukushima, Radioactivity has Seriously Harmed Wildlife," The Conversation, April 25, 2016; http://theconversation.com/at-chernobyl-and-fukushima-radioactivity-has-seriously-harmed-wildlife-57030

[8]. "Catastrophic Harm," International Campaign to Abolish Nuclear Weapons, http://www.icanw.org/the-facts/catastrophic-harm/

[9]. "Climate Disruption and Nuclear Famine," http://www.icanw.org/the-facts/catastrophic-harm/climate-disruption-and-nuclear-famine/

[10]. "Nuclear Famine: A Billion People at Risk," http://www.icanw.org/wp-content/uploads/2012/11/nuclear-famine-ippnw-0412.pdf

[11]. "Bengal Famine of 1943," https://en.wikipedia.org/wiki/Bengal_famine_of_1943#Famine.2C_disease.2C_and_the_death_toll

[12]. "Effects of Nuclear War on Health and Health Services," World Health Organization, Geneva, 1987; http://apps.who.int/iris/bitstream/10665/39199/1/9241561092_(p1-p82).pdf

[13]. "Consequences of a Large Nuclear War," Nuclear Darkness, Climate Change, and Nuclear Famine. http://www.nucleardarkness.org/warconsequences/hundredfiftytonessmoke/

[14]. A. Thompson, "Regional Nuclear War Would Affect Entire Globe," April 6, 2008; http://www.livescience.com/7479-regional-nuclear-war-affect-entire-globe.html

[15]. "Nuclear Weapon Yield," https://en.wikipedia.org/wiki/Nuclear_weapon_yield

INDEX

accidents,

    hunting 41-44

agriculture 110, 113, 115, 130, 139, 144

Alaska 5, 34, 105, 107, 135

algae 57, 76, 81

algal blooms 76, 81, 85

albedo 95

Al Qaeda 59, 62

Al Shabaab 19, 62

antarctic.. 90, 96, 100, 107-108, 135

aquaculture 70, 77, 81, 84, 115

arctic..96, 99, 106, 135

Asia.............i, 17, 19-21, 25-26, 30, 39, 64, 70, 107, 120, 125, 143, 150

atmospheric

    temperature 90, 97, 109, 120, 132

    rivers 111

Australia.. v, vi, 119-120, 124

bacteria 15, 45, 56-58, 75-76, 80, 84-85, 114, 124-127, 129, 136, 143

balance

    of nature .... v, 5, 7, 14-15, 23, 35, 49, 53, 146

bark beetle...101, 103, 136

bats .... vi, 124, 143

bears ....8, 25-26, 34, 37-38, 40, 66, 109 bends 70

bile..25-26

bioaccumulation 81-83

biodiversity ....vi, 23, 55, 59, 96, 115, 131, 149

birds...vi, 12, 37, 119, 124, 143

Boko Haram 63

bonobos..17, 125

bubonic.. vi, 125

buck laws 47, 49, 145

bugs.. vii, viii, 56

bullhooks... iv

bushmeat 2, 17-18, 22, 25, 62, 64, 122, 125, 143

cancer 20, 56, 58, 79, 82-83, 116, 124, 127, 129, 139-140

canned hunts...4, 27

carbon...................................... 88, 91, 102-106, 109, 113, 115, 144

carrying
    capacity 49, 50, 53-55, 118, 150

cascade ....v, 59, 99

cats .... vi, 32, 36, 65, 100, 125-126, 146

cattle........................... 38, 101, 103, 111, 113-115, 120-121, 125-126

Cecil 2, 25, 30, 70

chemicals.................................... 54, 59, 80, 82, 115, 130, 139-140

chikungunya... vi, 127

children ....viii, 2, 6, 8, 10, 38, 41, 44, 62-64, 67, 72, 79

chimpanzees.................................iii, 17, 21-22, 40, 62, 64, 125, 150

China..........17, 19-22, 25-26, 40, 88, 92, 107, 109-110, 116, 138-139

chipmunks... 11, 46-47, 125

cholera.................................. 75, 121, 124-125, 139

ciguatera.. 81-82, 85

CITES 16, 20, 23

climate change ....viii, 24, 85, 86ff, 144 collisions 1, 12, 48-50, 52-53, 142

Congress.. 12, 14, 26, 30, 33-34, 88, 91

corals .... iv, v, 58, 74-75, 85, 103, 123, 127, 132-133, 135-136

corruption 21, 61

cost 23, 48, 86, 89, 105-106, 109, 111-113, 121, 134, 142

cougars 13, 28, 37

coyotes .... vi, vii, 2, 4, 5, 7-8, 12, 14-15, 37, 42, 126

crop yields 110, 114

cruelty .... iii, v, 2-3, 6, 13, 22, 24-26, 28-29, 60, 66, 70, 149-150

crush videos iii

cyanide 12, 14, 74

deforestation 55-56, 103, 115-116, 122, 130, 136, 144-145

dengue .... vi, 125, 127

denning 12

desert 98, 110, 113, 115, 119, 130, 135

diet

   animal 57-58, 128

   human 79, 115, 128-129, 143

disease 55, 74-75, 77, 84, 105, 121, 124-126, 129, 136, 139, 143

dogs 11-15, 18, 32, 36, 38, 42, 65, 100, 120, 122, 126

dolphins 24, 75, 77

droughts 109, 111-114, 120, 127, 135

eagles 11-13, 32, 36-37, 65

ebola 18, 21, 45, 124-125

ecology .... vi, vii, 5, 23, 55, 146

ecosystem 23-24, 35, 57-59, 77, 87-88, 115, 123-124, 140, 143

efficiency 104, 144, 147

elephants ...... i, iv, 2, 18-20, 28, 40, 58-59, 61-62, 64, 70, 126, 145, 150

endangered

   species .... 5, 16-17, 26, 28, 30, 32, 37, 39, 65, 115, 123, 146 ethics 116, 149

extinction......17, 19-20, 23, 34, 36-37, 40, 53, 59, 74, 77, 100, 109, 115, 122, 136, 140

extreme
   weather................. 100, 106, 109-110, 112-113, 119, 121-123, 135

factory farming.................................... ii, iv, 71, 84, 122, 128-129

farm animals........................... ii, 3, 71, 100-101, 115, 120, 128, 143

farms
   lion, tiger...21, 25, 30, 40, 150

feedback... 15, 95-96, 99, 102, 109

finning... 22-23, 70, 143, 146

fish..69ff
   games 73
   learning 72, 78
   pain...72, 76
   sentience...72
   tools..72

floods....87, 89, 96, 107, 109-114, 116-117, 119, 121-122, 125-126, 134-136, 145

fossil fuels.........59, 87-88, 92, 94-95, 100-103, 108-110, 123, 144, 146

fox ....vi, vii, 4, 7-8, 15, 37, 67

fungus.. viii, 56-57, 76, 84, 114

gene pool..5, 35, 40, 85

gestation 26, 128

glaciers..89, 95-96, 107-108, 135

gorillas............................... 17-18, 21-22, 40, 59, 61-62, 64, 124-125

greenhouse
   effect... 86, 90, 101-102, 117
   gases........ 58, 87-89, 92, 95-96, 101-104, 109-110, 115, 130-131, 133, 145

Greenland 96, 99, 107-108, 135

growing season 96, 138-140

habitat .... ii, 15, 17-18, 21, 25, 33, 35-37, 40, 49-50, 57, 59, 105, 115, 117-119, 121-123, 136

heat index 98

heat waves 96-98, 100-101, 109-110, 112, 114-120, 122, 127, 131-132, 135-136, 145

HIV 125

Hong Kong 17, 19-20, 108

horses .... ii, iii, 33, 38-39, 44, 71, 120-121, 125-126, 128, 145

humane ed. 6, 146

hunting

    accidents 41-44

hurricanes 109-110, 112-113, 117, 119, 121-122, 135

insects 33, 56-58, 73, 114, 143

icebergs 90, 97, 107

ice sheets 99-100, 107-108

invaders 12, 36-37

invertebrates .... v, vii, viii, 33, 40, 55-59, 72, 118, 121, 123, 141, 143

ISIS 63-64

ivory 17-19, 62-64, 143, 146

IWC .... vii, 16, 23-24

Janjaweed 63

Japan 17, 20, 22-25, 76, 129

killing contests ....iii, 2, 7-8, 145

leghold traps 12-13, 20, 22, 34, 65-68, 143, 146

lions 2, 5, 25, 28-30, 38, 59, 61, 64, 70

livestock 5, 8, 11, 15, 113-115, 120-122, 126, 138
lobsters 69, 76-77
longevity 49, 124, 129
LRA..63
Lyme disease................................... vii, 2, 15, 45-47, 124, 127, 142

malaria.. vi, 56, 124-125, 127, 139
mantrap 66-67
meat 18, 20, 22, 24, 40, 62, 71, 83, 128-130, 144-145, 147
megafloods...111
mercury 24, 82-83, 85, 91
methane 96, 102-103, 109, 115, 145
models.. 97, 111, 138-139
monetary 105, 109, 111-112, 121
mortality...76, 84, 129
mosquitoes 58, 101, 127, 136
M44 ....12, 146

National Parks..21, 34
national security 17, 92, 129
NRA 19, 26, 30, 34
nuclear
    war.. iv, 93, 137-141, 145
    winter 138
nutrition..79, 114

oceans 23, 55, 58, 70-71, 73, 77, 80, 82, 90, 95, 97, 101, 103, 106,
            132-133, 135, 141, 150
ocean temp. 74, 90, 97
ocean currents 96, 98, 108
octopus 69, 72, 76

organized
  crime 17, 60, 62
ozone 103, 136, 139, 141

park rangers... 18, 60-62, 64, 143
pathogens 56, 75, 80-81, 84-85, 124, 126-129, 136
permafrost 95-96, 102, 135
pesticides.. vi, 56-58, 82, 84, 115, 130
plague.. vi, 56, 124-125, 127
plankton...58, 77, 81, 85, 140
poaching......i, iii, iv, 2, 16-21, 23-26, 34-35, 37, 40, 59-62, 64, 143
poisons 12-14, 20, 22, 62, 81
polar bears...5, 40, 112, 123
politics...v, 31, 93, 103
pollution 40, 54-55, 73, 80, 122-123, 135, 139
population 5, 55, 73, 131
predators............v, vi, vii, 2, 4-5, 7-10, 12-15, 28-29, 34-38, 47, 49,
             53, 58, 82, 146
prey ....vii, 15, 25, 35-36, 69, 72-73, 85, 109

rabbits.. iv, 2, 4, 7-9, 36
rabies..12, 124
rainforests...54, 111, 113
reefs...................vi, 55, 58, 72, 74, 77, 103, 123-124, 132-133, 136
refugees..32, 63, 92, 110, 117, 119, 136
refuges.. 31-35, 65, 68, 143, 146, 148
rhinos..17, 20, 28, 37, 40, 59, 61, 64, 143
Russia.. 17, 20, 31-32, 98, 138, 140

Safari club 5-6, 26, 30, 34
sanctuary .... vii, 24, 31-32
schools...6, 63, 146

scombroid..81

sea ice...96, 99, 106, 109, 122

sea level.............85, 87-89, 96-97, 100, 106-108, 111, 116-117, 122,
       134-135

sharks .... vi, 22-23, 40

SIV 124-125

smog..136

smoke 94, 105, 110, 134, 136, 139, 140

snares.. 12-14, 18, 20, 62, 65

starfish..v, 55

storms.. 109-110, 112, 127

storm surges .... vi, 106, 121

suffering .... i, ii, vii, 2, 4, 6, 9, 13, 27-29, 47, 50, 65, 68-71, 76,
       148-150

temp. index..98

temporal bias...149

terrorism................................ 17, 19, 54, 61-64, 129, 146

tigers.......... i, 2, 17, 20-21, 25, 28-30, 35, 37, 39-40, 58-59, 61, 143

ticks.............................vi, vii, 15, 45-47, 126-127, 136, 143

toxins.. 76, 80-82, 84, 91

trapping 5-6, 31-33, 35, 65, 148

trawlers.. 70-73

tuberculosis 55, 124, 126

urchins.. vii, 55, 59

vegetarian..129, 144

violence,..v, 3, 63

  human..v, 6, 149

viruses 15, 75-76, 80, 84, 124-127

West Nile 101, 124-125

whales .... vii, 23, 37, 39-40, 58-59, 73, 77, 82, 123

whaling .... vii, 23-24, 143

wildfires 59, 97, 103-105, 109, 117-118, 120, 122, 134, 136, 145

wildlife

    agencies 5-6, 13, 16, 47, 49-50, 65

    Services 11, 14

    trafficking 16, 25, 61, 64

    war on .... vi, 142, 148

wolves 8, 15, 26, 37-38, 59

Zika .... vi, 124

zoos 28-29, 36, 150